# NETYMOLOGY

# NETYMOLOGY

From Apps to Zombies: A Linguistic
Celebration of the Digital World

## Tom Chatfield

Quercus

First published in Great Britain in 2013 by
Quercus
55 Baker Street
Seventh Floor, South Block
London
W1U 8EW

A CIP catalogue record for this book is available
from the British Library

Hardback ISBN 978 1 78087 993 2
Ebook ISBN 978 1 78087 994 9


10 9 8 7 6 5 4 3 2 1

Text and plates designed and typeset by Ellipsis Digital Ltd

Printed and bound in Great Britain by Clays Ltd. St Ives plc

# Contents

Introduction    1

1. Computers    7
2. Signs of our times    10
3. Marking up    13
4. The burning power of a name    16
5. Myths and monsters    19
6. Speak, memory    22
7. Why wiki?    25
8. Buffed-up gamers    27
9. Very, very big and very, very small    30
10. The names of domains    33
11. Rise of the robots    36
12. Cyber-everything    39
13. Three-letter words    42
14. Everyone's an avatar    45
15. On memes    48
16. Hacking through the net    51
17. Do you grok it?    54
18. Sock puppets and astroturf    57
19. Bluetooth    60
20. The Cupertino effect    63
21. The Scunthorpe problem    66
22. The coming of the geeks    69
23. Beware of the troll    72

24. Bitten by bugs    75

25. Bits, bytes and other delights    78

26. Twinks, twinked and twinking    81

27. Talking less about trees    84

28. ZOMGs, LOLZ    87

29. Happy pew pew to you    90

30. Lifehacking    93

31. The multitasking illusion    96

32. The Streisand effect    99

33. Acute cyberchondria    102

34. Casting the media net    105

35. Bionic beings and better    108

36. Technological black holes    111

37. Google and very big numbers    114

38. Status anxiety    117

39. The zombie computing apocalypse    120

40. To pwn and be pwned    123

41. Learning to speak l33t    125

42. Emoticons    128

43. Getting cyber-sexy    131

44. Slacktivism and the pajamahadeen    134

45. Gamification and the art of persuasion    136

46. Sousveillance    139

47. Phishing, phreaking and phriends    142

48. Spamming for victory    145

49. Gurus and evangelists    148

50. CamelCase    151

51. The Blogosphere and Twitterverse    154

52. Phat loot and in-game grinding    157

53. Meta-  160

54. TL;DR  163

55. Apps  166

56. Fanboys and girls  169

57. Welcome to the Guild  172

58. Facepalms and *acting out*  175

59. Finding work as a mechanical Turk  178

60. Geocaching  181

61. The beasts of Baidu  184

62. Snowclones  187

63. Typosquatting  190

64. Egosurfing and Googlegangers  193

65. Infovores, digerati and hikikomori  195

66. Planking, owling and horsemanning  198

67. Unfriend, unfavourite (and friends)  201

68. Sneakernets and meatspace  204

69. Going viral  207

70. Dyson spheres and digital dreams  210

71. Welcome to teh interwebs  213

72. On good authority  216

73. A world of hardware  219

74. Darknets, mysterious onions and bitcoins  222

75. Nets, webs and capital letters  225

76. Praying to Isidore and tweeting the Pope  228

77. QWERTY and Dvorak  231

78. Apples are the only fruit  234

79. Eponymous branding  237

80. Mice, mouses and grafacons  240

81. Meh  243

82. Learn Olbanian!   245

83. Booting and rebooting   248

84. Cookie monsters   250

85. Going digitally native   253

86. Netiquette and netizens   256

87. The names of the games   259

88. Flash crowds, mobs, and the slashdot effect   262

89. Godwin's law   265

90. From Beta to Alpha to Golden Master   268

91. Mothers and daughters, masters and slaves   271

92. Bit rot   274

93. Non-printing characters   277

94. Wise web wizards   279

95. Disk drives   282

96. Easter eggs   285

97. Microsoft family names   288

98. Why digital?   291

99. Filing away our data   294

100. Artificial Intelligence and Turing tests   297

. . . and finally   300

*Acknowledgements*   302

*Select bibliography and further reading*   304

*Notes and references*   307

# Introduction

I've never understood those who lament the internet as a kind of death for the English language. Just look at the ways in which most of us, every day, use computers, mobile phones, websites, email and social networks. Vast volumes of mixed media may surround us, from music to games and videos. Yet almost all of our online actions still begin and end with writing: text messages, status updates, typed search queries, screens packed with verbal exchanges, and underpinning it all countless billions of words.

The twenty-first century is a hypertextual arena in several senses (*hyper*, ancient Greek: 'over, beyond, overmuch, above measure'). Digital words are interconnected by active links, as they can never be on the page. But they are also above measure in their supply, their distribution, and the range of roles they play – from casual registers unthinkable a century ago to the most elaborately scholarly of debates and exegeses.

New things have always required new words, of course, and technological developments have fed into both written and spoken language since the earliest times. This is considerably trickier in some languages than others, thanks to the difficulty of expressing entirely novel words within fixed systems like symbolic characters (a computer in Mandarin is, for example, a *diannao* – 电脑 – which literally means an 'electric brain'; while *dian*, meaning 'electric', itself originally described 'lightning').

With English, however, we have not only a language that has spent the best part of a millennium gleefully adopting new terms and ideas; we also have something truly international, whose history has become part of the history of countless places, peoples and movements.

Today, globalization and new technology have vastly accelerated both the speed and the scale of linguistic evolution – a process that is blurring many boundaries between languages, dialects and registers to the point of disintegration. Perhaps the greatest difference between digital and pre-digital times, however, is that it's now a written (or, more precisely, a typed) rather than a spoken language driving these changes. The future of written words lies onscreen – and these screens are steadily transforming not only how we communicate, but what we mean and think.

Separated from human voices and faces, new conventions and registers are developing to express the emotional tone of typed words: from smiley faces built out of punctuation marks to subcultures of mockery and praise that adhere, in their own way, to unwritten etiquettes as elaborate as those of any Tudor court. Separated from pens and pages, too, these words are active agents in the world in a new sense: tools that can be countlessly replicated, adapted and shared.

For some people, this loosening and cheapening of words is a tragedy, dragging cultural standards ever closer to the gutter. For others, though – and I count myself among them – what's happening is simply too big, too important and too exciting either to summarize or condemn.

Standard English, in which I'm typing these words, appeared over several centuries as the dominant form of 'proper' writing in our language. It was an enterprise that, for the first time, allowed the words we use to be regulated by reference to central authorities – dictionaries, grammars, experts. The fruits of this standardization include a truly global written culture and a clear, comprehensible language of official culture and organization.

Whether we like it or not, however, many of the official intentions behind Standard English are already unofficially defunct. For the first time in history, we live in a culture not only of mass literacy (itself a relatively recent revolution), but of mass participation in written discourse. Online, reading and writing – which not so long ago were among the most costly and elite of human activities – are almost infinitely available at little or no cost. For better and for worse, we are no longer simply speakers of our own tongue: we are all becoming both authors and audiences.

Indeed, the art and science of computer programming has brought with it an entirely new species of language:

a form of written expression whose terms encode not only meanings but also entire interactive systems. In an age of information technology, information itself is becoming the stuff of self-sustaining worlds — worlds that consume an increasingly substantial proportion of our attention, innovation, effort and desire for self-expression.

Sometimes it can all be overwhelming, with any attempt to describe the changes facing us almost instantly out-dated. Indeed, perhaps the only and best solution to this situation is the internet itself: an eternally unfinished collaboration, pooling the words of many millions.

Writing and researching this book, I have been both daunted and bewildered at times. Above all, though, I have felt exhilarated — and lucky as a lover of language to be alive today. Many of the great scholars, sages and authors of the past would have given anything to possess what a single screen can offer me in an instant: tens of millions of books, hundreds of millions of pages, and over a billion human voices competing in articulate cacophony.

These new stories of our language can, I believe, conjure the fragmentary story of our present with a vividness appropriate to the experience of living it. For each etymology isn't so much a complete tale as an ongoing negotiation — a balance of meanings and readings for which there can be few better images than the restless texture of the digital world itself.

Clarity, concision and comprehension have always been the keys to using language well — something no amount of technology can change. I hope this book helps you in cultivating these, together with that most important 'c' of all: curiosity, without which we might as well not bother speaking at all.

# 1.

# Computers

It's easy to forget that, for most of its existence, the English word 'computer' referred not to machines, but to people who performed calculations. First used in the seventeenth century, the term arrived via French from the Latin *computare*, meaning to count or add up. *Computare* itself derived from the combination of the words *com*, meaning 'with', and *putare*, which originally meant 'to prune' in the sense of trimming something down to size, and which came to imply 'reckoning' by analogy with mentally pruning something down to a manageable estimate.

Long before eminent Victorians like Charles Babbage had even dreamed of calculating machines, human computing had been vital to such feats as the ancient Egyptians' understanding of the motion of the stars and planets, with mathematicians like Ptolemy laboriously determining their paths (he also managed to calculate pi accurately to the equivalent of three decimal places: no mean feat for the first century AD).

As mathematics developed, the opportunities for elaborate and useful calculations increased — not least through the development of tables of logarithms, the first of which were compiled by English mathematician Henry Briggs in 1617. Such tables immensely simplified the complex calculations vital to tasks like navigation and astronomy by providing pre-calculated lists of the ratios between different large numbers – but whose construction required immense feats of human calculation both by mathematicians and increasingly necessary groups of trained assistants.

Even as recently as the Second World War, when Alan Turing and his fellows were establishing the revolutionary foundations of modern computing, the word 'computers' still referred to dedicated human teams of experts – like those working around Turing at Bletchley Park in England.

According to the *Oxford English Dictionary*, it wasn't until 1946 that the word 'computer' itself was used to refer to an 'automatic electronic device'. This was, of course, only the beginning; and since then both the sense and the compound forms of the word have multiplied vastly. From 'microcomputers' to 'personal computers' and, more recently, 'tablet computers', we live in an age defined by Turing's digital children.

It's important to remember, though, just how recently machines surpassed men and women in the computation stakes. As late as the 1960s, teams of hundreds of trained human computers housed in dedicated offices were still being used to produce tables of numbers: a procedure that the first half of the twentieth century saw honed to a fine art, with leading mathematicians specializing in breaking down complex problems into easily repeatable steps.

It's a sign of how fast and entirely times have changed since then that human computation is almost forgotten. And yet, in different forms, its principles remain alive in the twenty-first century – not least under the young banner of 'crowdsourcing', a word coined in 2006 in an article for *Wired* magazine by writer Jeff Howe to describe the outsourcing of a task to a large, scattered group of people.[1]

From identifying the contents of complex photographs to answering fuzzy questions or identifying poorly printed words, there remain plenty of tasks in a digital age that people are still better at than electronic computers. We may not call it 'human computation' any more, but the tactical deployment of massed brainpower to solve some problems remains more potent than ever.

# 2.
# Signs of our times

Until its first use in emails in 1971, the @ symbol was an obscure object used to indicate pricing levels in accounting. Since then – when it was selected by programmer Ray Tomlinson, from around a dozen available punctuation characters, for use in his brand new e-mail software system – it has become one of the world's most widely used symbols, and has gathered a bewildering and colourful variety of different descriptions in different languages.

While in English it is simply called the 'at sign', others are more poetic. In Italy it is *chiocciola*, 'the snail', thanks to its shape, a phrase the French also echo with *petit escargot*. In Finnish it is thought to look more like a curled-up cat (*miukumauku*) while Russian leans towards a dog (*sobaka*). The Chinese sometimes call it *xiao laoshu*, or 'little mouse'. But perhaps most colourful of all is the German interpretation: *klammeraffe*, or 'spider-monkey'.

Still more eccentric is the story of Apple's 'command' key, marked by a square with looped corners, or ⌘. Known properly as the St John's Arms, it's an ancient, knot-like heraldic symbol, dating back in Scandinavia to at least 1000 BC, where it was used to ward off evil spirits and bad luck.

It's still found today on Swedish maps, representing places of historical interest, thanks to its (approximate) resemblance to the tower of a castle viewed from above. Anecdote has it that the sign was suggested at Apple by the graphic designer Susan Kare, after Steve Jobs declared in a 1983 meeting that using Apple's own brand symbol all over its keyboard was an excessive example of 'taking the Apple logo in vain'.[2] To many modern Mac users, though, it's simply 'the command squiggle' or 'splodge'.

Spare a thought, finally, for the # symbol and the unlikely diversity of English names it carries. If you're using a social media service like Twitter, you'll probably call it a 'hash tag'; if you're pressing the button bearing this symbol on a phone in North America, you should probably call it the 'pound sign'; and if you're trying to be internationally unambiguous, you might call it the 'number sign'. If you're a proof-reader, meanwhile, making this mark on a manuscript indicates the need to insert a space. You might also be tempted to call # a 'sharp sign', in the musical sense – but you'd technically be incorrect to do so, as the lines are angled differently if musical notation is intended.

What, then, is the origin of #? A far younger symbol than St John's Arms and its ilk, one of its first recorded uses dates from 1920s America, when it served as a short-hand for 'pounds' in the sense of weight. Its subsequent adoption by telephone engineers at Bell Labs in the 1960s as a generic function key on their new 'touch tone' tele-phones helped enshrine this name in the popular American consciousness.

If, however, you're looking to sound really clever, you can always call # an 'octothorpe' – a word fabricated especially by Bell engineers to match the symbol on their new phones. 'Octothorpe' mixes the Greek word for eight, *octo* (describing the figure's eight points: one on each end of its four lines) with the mysterious ending 'thorpe', which may or may not refer to American athlete Jim Thorpe or the Old English word for a village, *thorp*. Thus far, the name has failed to catch on.

# 3.

# Marking up

Behind the scenes of the very latest versions of our written language, some intriguingly ancient terms have found new homes. Hypertext Markup Language, or HTML, is the bread and butter of the world wide web. The term 'hypertext' itself was coined as early as 1963 by the American sociologist Ted Nelson;[3] but even this pales in comparison to both the word 'markup' and many of the most common terms in online markup languages, which date back not to the first days of digital technology, but to a far earlier transformation: the birth of printing.

Printing with movable type first appeared in Europe in the fifteenth century, and was a laborious process that usually involved hand-written manuscripts being 'marked up' with instructions to the printer as to how they should be presented on the page: which words should be in bold, italics, headings, underlined, or set out separately from the main text.

Several of these printer's terms survive to this day online: from the abbreviation 'em' signalling 'emphasis' (*type in italics*) to the use of the tag 'strong' to **signal bold type.** The 'chevron' style of bracket within which these terms are enclosed in HTML – '<' and '>' – is, meanwhile, even older than printing, with a name first coined in the four-teenth century based on its apparent resemblance to the rafters of a roof (*chevron* in Old French).

There's also a pleasing physicality to many of the behind-the-scenes labels of the modern web. Consider the standard differentiation of verbal elements on a page into the 'head' and 'body' of a text, for example – a metaphorical divi-sion as ancient as they come.

That HTML is based on English words is a historical accident – because its inventor, the creator of the world wide web, was the English computer scientist Tim Berners-Lee (even though he was in fact working for CERN in Geneva when he created the web in 1990). And one con-sequence of the existence of truly global standards like HTML is the universal application of their terms. No matter what country or language a website is based in, the markup terms within it remain the same: that is, English words like 'head' and 'body' will remain silently present within the encoding of a page, telling every web browser in the world how it ought to look.

Since its first specification in 1990, which contained just eighteen different kinds of digital 'tags' – itself a 600-year-old English word of uncertain origin, and which originally referred simply to a 'small hanging piece' of something – HTML and its offspring have grown vastly in complexity. They consist largely of terms devised to ensure maximum clarity and simplicity, from 'object' to define any embedded object to 'button' to define a clickable button.

There are still etymological riches to be unearthed, however. Even that most familiar of typographical terms, 'font', carries a half-millennium of history with it, deriving ultimately from the Middle French word *fondre*, 'to melt', thanks to the sixteenth-century need to melt down lead in order to make casts of letters for early printing.

Technological times may change fast – but words have their own momentum.

# 4.
# The burning power of a name

The best brands become icons: the VW Beetle for the 1960s; the iPod for the noughties. When it comes to the twenty-first century, the book trade has been lumbered with something equally significant: a brand that may live up to its name by sending the old order up in smoke.

The brand in question is Amazon's electronic book reader, the Kindle – a device that reached its fifth generation in 2012, and that now accounts for more e-book sales through Amazon than the site shifts in conventional paperbacks.

The word 'kindle' itself means to set alight, taken from the Old Norse *kynda*, and is defined on the home screen of Amazon's device as both to 'light or set fire' and, more poetically, to 'arouse or inspire'. Delightfully – and presumably coincidentally, where electronic books are concerned – 'kindle' is also the collective noun for a group of kittens, thanks to the Middle English word *kindel* meaning 'to give birth', itself a distant relative of the Norse original.

Most significantly when it comes to literature, the word

## The burning power of a name

'kindle' features in a famous epigram from the French enlightenment thinker Voltaire in praise of book-learning: 'we fetch it from our neighbours, kindle it at home, communicate it to others, and it becomes the property of all.' The sentiment itself is from Voltaire's 1733 *Lettres Philosophiques* – originally written in English – and in particular from the last passage in his letter on 'That Singular Poem Called "Hudibras"; and Dean Swift'.[4]

Voltaire's work has a complex publication history, and it's seemingly only thanks to a rather loose 1786 retranslation of the French version that we enjoy the epigram as it is today.[5] This is probably just as well, as the French verb *allumer* (from the Latin *illuminare*, 'to light up') is rather less striking than its northern European cousin, and an 'Amazon Allumer' doesn't have the same ring.

The celebration of written words is all very well but, to some traditional publishers' ears, 'kindle' may have a distinct ring of Ray Bradbury's *Fahrenheit 451* to it, with Amazon's 'firemen' gleefully building a funeral pyre for the printed word.

It's a theory that won't have been dampened by the company's decision to name the latest incarnation of its reader the 'Kindle Fire' – two words that, spoken in the right tone, can sound more like an incendiary command than a brand name.

In any case, the firm is more than aware of its revolutionary heritage. Amazon's 2011 advertising campaign for the device featured Voltaire's original line prominently, followed by the promise that 'from Kindle, fire is born' — grammatically dubious, but suitably unambiguous in its ambition.

Contrast this to the Kindle's main rivals, and you'll find a stark divergence of naming ferocity. Barnes and Noble's Nook is simply the word 'book' with a different first letter; Toronto-based company Kobo's eponymous eReader, the Kobo, is 'book' with its letters rearranged; while Sony's offering is known only as the Sony Reader.

Voltaire it isn't. Given that Apple's rival software to Kindle is called iBooks, and Google's called Google Books, you'd be forgiven for thinking that only one company working with onscreen words has any fire in its belly.

# 5.

# Myths and monsters

Bestiaries — literally 'books of beasts' — were among the most exquisite written products of the Middle Ages. Lavishly illustrated guides to the animal kingdom, they melded fact and rumour to create mystically charged taxonomies of nature. The whale, or leviathan, was a Satanic creature whose belly represented hell on earth to those whom it swallowed, like Jonah. The bear gave birth to shapeless, eyeless lumps of flesh, then licked them into cubs.

As well as bestiaries' beauty and spiritual force, their pre-scientific systems of correspondence helped to make a seemingly limitless world comprehensible. Today, in what may seem to be the most rational of realms, we're seeing a strange revival of this kind of magical thinking and language.

Take the terms we turn to when things start to go wrong with our bright new machines. Viruses, worms, infections, Trojans — the flavour is more medico-mythological than rationally pristine. And its subtexts are viscerally evocative.

A 'Trojan', for instance, takes its name from the *Aeneid*'s Trojan horse, used by treacherous Greeks to gain entry to the city of Troy within a supposed gift, and is a malicious program lurking inside a benign-seeming exterior. More prosaically, 'worms' are self-replicating pieces of malicious software that wriggle their way through the internet's underbelly; then there are computer viruses which, like their biological namesakes, can only reproduce and spread when they are parasitically attached to an existing file or program.

On the other side, meanwhile, we find equally evocative words. Since 1963, useful programs that run in the background rather than under a user's direct control have been known as 'daemons'. The term, an alternative spelling of 'demon', dates back to spirits found in Greek mythology. But the particular daemon the programmers who coined the term while working at MIT had in mind was a more modern kind of myth: Maxwell's demon, an entity invented as a thought experiment in 1867 by the Scottish physicist James Maxwell.

Maxwell imagined his demon using its superhuman powers to move individual molecules around within a container, causing them to violate the second law of thermodynamics. As explained by MIT's Professor Fernando Corbato (in response to an etymological trivia column in *The Austin Chronicle*, no less): 'Maxwell's daemon was an imaginary agent which helped sort molecules of different speeds and worked tirelessly in the background. We fancifully began to use the word daemon to describe background processes which worked tirelessly to perform system chores.'[6]

Modern computer daemons tend to look after scheduled tasks on networks, answer and redirect emails automatically, or help configure hardware — hardly the stuff of myths. Their ghostly presence within our machines, however, is itself something of a miracle: the arbiters of self-contained worlds, seamlessly ticking over.

# 6.
# Speak, memory

The words we use say more about us than we usually realize. In a sense, they also use us – and never more so than when we're speaking about what it feels like to *be* us.

Take an innocuous human term like 'memory'. The word itself has been with us in English for a good 800 years, arriving from Latin via French (*memoria* and *memorie* respectively) in the mid-thirteenth century, with little essential change in its significance for several thousand years.

In 1946, however, memory stopped being a strictly biological business, when it was applied for the first time in history to the memory of something inanimate – an early electronic computer.

Today, using 'memory' to describe the physical microchips encoding a computer's files is as familiar a usage as describing the human process of remembering. It's also an implicit analogy that has had a significant impact on the way we think about ourselves.

# Speak, memory

Over the last half-century, computers have become a dominant metaphor for the way we describe our own minds. From talk of processes and calculations to belief in nodes, routines, modules and components, accounts of our brains no longer feature the homunculi of early twentieth-century illustrations, or the metaphysical humours and passions of classical thought. Instead, we turn to the hard drives and circuit boards of digital machines for analogy.

Consider what it means for a machine's memory to function well. It should be large, free from errors, rapid, searchable, easy to expand or wipe clean, and categorized into comprehensive and unambiguous sections. The bigger, cleaner, brighter and faster it is, the better.

Speak of memory today and these are some of the associations that will be summoned, whatever the context. Yet they bear little resemblance to the architecture of a human mind – in which recall is serendipitous, embedded in a unique personal history, entwined with feelings, places and beliefs, and constantly shifted by the mind's churning present tense.

Just as steam-powered machinery left its metaphorical mark during the industrial revolution, the language we bring to bear on our own minds is increasingly shaped by computing: from talk of 'processing' and 'downloading' ideas to acts like 'rebooting' our attitudes, 'reprogramming' our thinking or even 'rewiring' our brains.

We seek to understand ourselves – as we must – with the words we have. In digital technology, we possess a unique kind of mirror for self-reflection; but also an analogy for intelligence to whose imperfections we must remain alert. Computers may help us to remember and to record; but there's a world of difference between mere forgetting and deletion, or a moment recalled and a moment merely recorded.

# 7.
# Why wiki?

Thanks to the world's single most successful repository of human knowledge, all of us by now know a little Hawaiian — at least indirectly. I'm referring to Wikipedia, which created its name by bolting the aptly rapid Hawaiian word for 'fast', *wiki*, onto the back half of 'encyclopaedia' — the Greek word *paideia* meaning 'education'.

There can be few better two-word descriptions for what Wikipedia does than 'quick education'. But there's more than this to the story, for 'wikis' constitute a far larger category of object than Wikipedia itself — and one boasting an intriguing, if brief, etymological history.

Wikipedia was founded in 2001, and today boasts almost four million articles in English — and as many again in over a hundred other languages. But the idea of a 'wiki' goes back to the near-Neolithic web date of 1995, and to American programmer Howard G. 'Ward' Cunningham's vision of a website whose users could rapidly edit all of its pages.

Cunningham dubbed his brainchild WikiWikiWeb. As he subsequently explained in correspondence with an etymologist from Oxford University Press, intrigued at being able directly to quiz a new word's creator, 'I chose the word wiki knowing that it meant quick. I also knew that in Hawaiian words were doubled for emphasis. That is, I knew that wiki wiki meant very quick . . . I was not trying to duplicate any existing medium, like mail, so I didn't want a name like electronic mail (email) for my work.'[7]

Since then, Cunningham's model of communal website building has become a template for millions of sites. It has also, among other things, inspired the unlikely tribute of its own 'backronym'. That is, an alleged acronym retrospectively derived from its name: 'What I Know Is.'

Backronyms abound online, other (in)famous examples being the retro-conversion of Yahoo! into 'Yet Another Hierarchical Officious Oracle' and Microsoft's search engine Bing into 'Because It's Not Google'.

Yet perhaps the finest historical pedigree for misleading etymologies in technical circles belongs to the notion that SOS stands for 'Save Our Ships' or 'Save Our Souls'. In fact, it was adopted as a distress signal in 1905 simply because of the ease of sending these letters in Morse code via the then-cutting-edge electric telegraph system.

# 8.

# Buffed-up gamers

I've been a video games fan for most of my life, and grew up within the dialect of the tribe – a space especially rich in terms (and boasts, and taunts) for winning and losing, and for describing precisely how much better or worse various Elvish enchantments make your character.

When something boosts your powers or status in a game, you talk about getting 'buffed' or acquiring a 'buff'. As you might expect, the term derives from the idea of 'buffing up' something to improve its appearance. The phrase originated in English in the late nineteenth century, referring to the practice of using a 'buff' or piece of leather for polishing.

A leather 'buff' itself got its English name in the 1570s from the term 'buffe leather', which originally described buffalo hide – making in-game talk of having a 'good buff' equivalent, etymologically at least, to admiring the excellence of a large cow.

The idea of buffing up is easy enough to understand outside of any digital context. Far more esoteric, though, is the term used to describe its opposite: having one's powers reduced, or being 'nerfed'.

A pleasantly onomatopoeic word – with hints of both 'nerd' and 'worse' – nerf is also a young term, tracing its roots to the cult 1997 game Ultima Online. One of the first true massively multiplayer online games, in which tens of thousands of users collaborated and competed together in a medieval fantasy world, Ultima Online also boasted an extremely vocal player community ready to voice loud dissent at any perceived injustice.

At one point in the game's history its designer, Raph Koster, hosted an online chat to discuss the relative ineffectiveness of weapons like swords in the game as compared to bows or magical spells. Koster promised he would look into the problem of what he called 'nerf swords' – a tongue-in-cheek reference to a popular brand of toy foam sword, whose name came from the acronym 'Non-Expanding Recreational Foam'.[8]

Unknowingly, Koster had gifted the gaming world a key term – and one that's now enshrined in its professional vocabulary, with game designers and companies regularly debating how to 'nerf' over-powered aspects of a game's mechanics.

Pleasingly, the nerf circle has also been squared in recent years by the release of several official Nerf-foam branded video games, complete with games console accessories that double as fully functional foam dart guns.

Meanwhile, on an almost-certainly unrelated note, the

term 'nerf' also refers in George Lucas's Star Wars universe to a species of alien herbivorous mammal, with 'nerf herder' serving as a fond insult during the course of *The Empire Strikes Back* – a pop cultural pedigree august enough to have had at least one band named in its honour.

# 9.
# Very, very big and very, very small

For the ancient Greeks, a 'myriad' was as big as it got. The word is used today to mean a large, undefined (and quite possibly uncountable) quantity of something – but in ancient Greece it referred precisely to the number 10,000, and was the largest single name for a number that existed.

You could multiply myriads, of course, and Archimedes referred to a 'myriad myriad' (100,000,000) as part of a work called *The Sand Reckoner*, in which he set out systematically to calculate how many grains of sand it would be possible to fit into the entire universe.

Archimedes was testing the number system of his era to its breaking point. The figure he eventually came up with for the size of the universe – which reckoned it as around two light years across, in modern measurements – was itself many billions of times too small. Even today, however, technology and human knowledge continue to push at the boundaries of numbers and language.

One recent word is the prefix 'exa': an addition to the official International System of Units for naming large quantities.[9] Most people are aware that 'megabytes' are units of computer storage on a scale of one followed by six zeros (millions), while 'gigabytes' are a scale of one followed by nine zeros (billions) and 'terabytes' are a scale of one followed by twelve zeros (trillions). Beyond this, however, the terms become increasingly obscure – and recent.

One followed by fifteen zeros has since 1975 had the official prefix 'peta', while in the same year the prefix 'exa' was established for a one followed by eighteen zeros – the largest officially named international number at that point in history. Both 'exa' and 'peta' are terms adapted from ancient Greek, in each case by removing a letter from one of the numbers between one and ten. 'Peta' is a shortened form of the Greek *penta*, meaning five, because it denotes five times as many zeros as the basic unit for large numbers, one thousand; 'exa' is a shortened form of the Greek *hexa*, meaning six, because it denotes six times as many zeros as one thousand.

'Exa' numbers exist on a scale so huge that the mere stuff of the world around us rarely comes close. Just one 'exasecond' is around 32 billion years, more than double the age of the universe. Yet by 1991, it was decided that new terms needed to be brought into use for still larger numbers: 'zetta' (a one followed by 21 zeros, based on the Greek for seven, *hepta* – because it has seven times as many zeros as one thousand – with the deliberate addition of a 'z' to avoid duplicating 's' as the first letter of a prefix) and 'yotta' (a one followed by 24 zeros, based on the Greek *okta*, eight, with the use of 'y' to avoid the potential confusion of using the letter 'o' – which could be read as a zero – to begin a prefix).

Vast though they may be, these scales are already starting to come into play thanks to both the level of our increasing knowledge about the universe and the sheer quantities of data we ourselves are generating through faster and more numerous computers. We don't yet measure our hard drives in 'yottabytes' – that is, 1,000,000,000,000,000,000,000,000 bytes of information – but it's not outlandish, if present trends continue, to think that some day we eventually may. We have already started to speak increasingly casually of storing terabytes (1,000,000,000,000 bytes) worth of information within home computers.

The future, of course, isn't just big – it's also incredibly small, not least in the scale of the technologies required to store all this information, and in the extremely small other things these technologies allow us to see. 'Yocto-' (a 1 preceded by 24 decimal places of zeros) is currently as small as it officially gets. In the years to come, though, things are only going to get more extreme in both directions.

# 10.
# The names of domains

The internet functions across the world thanks to what you can think of as a consistent internal grammar — a set of basic rules governing the organization and presentation of information, so that every part of the system is able to communicate successfully with every other part. One of the most fundamental and familiar parts of this grammar is the Domain Name System, or DNS.

It's this system that allows us humans to give recognizable names — from www.bbc.com to tomchatfield.net — to different online resources, not to mention organizing these resources into a strict, coherent global hierarchy. This system is also, however, a largely unsung treasure-house of some of the most iconic new linguistic formulations to enter our world in the last half-century.

The word 'domain' is a distinctly aristocratic one, entering English via Scottish in the fifteenth century from the Old French term *demaine*, meaning a lord's estate; itself a word drawn from the Latin *dominus*, meaning lord or master. Domain names were first described in 1983 and revert nicely to this classical type. For while the technology itself is essentially a mnemonic, converting the abstraction of pure numbers into words and letters, it represents one of modernity's most significant acts of naming: an almost universally accessible form of lordship, bringing with it mastery of one's very virtual estate.

Like much of the digital world, domain names exist primarily to be read and written (or typed) rather than spoken – one interesting side-effect of which has been to train recent generations to speak their punctuation out loud. 'Dot com' is such a standard turn of phrase it no longer seems odd, to the extent that it has now entered the language as a respectable noun for any online business.

Indeed, domains ending in 'dot com' have a kind of blue chip reliability to them that still makes them the most desirable class of domain, even though the '.com' ending is just one among twenty-one generic 'top level' domains. Others include the slightly esoteric '.info' and '.net' – and the recent explicit addition '.xxx', for adult-themed content.

Besides these generic domains, there are also some 250 top-level domains particular to different countries, one of which – '.tv', belonging to the tiny Polynesian island nation of Tuvalu – represents a substantial proportion of the value of the island's economy, thanks to its happy usefulness as the standard abbreviation for 'television'. The island's government rents out the right to supply .tv domain names to a private company for a lease worth millions of dollars.[10]

Spare a thought, finally, for those who have unintentionally fallen foul of the system. Once upon a time, for example, there was a service for helping you find a therapist called Therapist Finder, whose address www.therapistfinder could unfortunately be read in an entirely different sense.

Then there were the Mole Station Native Nursery, www.molestationnursery.com, eBay rival site Auctions Hit at www.auctionshit.com, and countless others — most of which no longer exist, save in the internet's infinite memory.

As a modern art, domain naming may not be up there with other forms of creative writing, but it's a vital verbal consideration nonetheless — and one most of us find ourselves reading, hearing or saying something about every single day.

# 11.
# Rise of the robots

Literature has always been a rich source of new words for English. Top of most lists of contributors are the great authors of the sixteenth and seventeenth centuries: William Shakespeare, Edmund Spenser, Andrew Marvell, John Milton and peers. In the two centuries after 1500, authors gifted the English language over 30,000 new words – largely from Latin and Greek, but also from French, Italian, German and even Arabic. It was a time of fertile collision between verbal worlds, matched by the great cultural and intellectual leaps forward that would later be dubbed the Renaissance.[11]

We live in an era of similar collisions, and while comparisons with Shakespeare and the Renaissance may seem presumptuous, we also have literature to thank for some of the most telling new words of our times.

One of the earliest among these arrived in 1920, thanks to the Czech writer Karel Čapek and his science fiction play *Rossum's Universal Robots* (despite being written in Czech, this English phrase was used as its official subtitle).

# Rise of the robots

Čapek was looking for a new word to describe artificial humans, and chose an adaptation of the Czech word *robota*, meaning 'serf labour' – an idea he later noted had come from his brother Josef.[12]

'Robots' were a speculative dream, but one with a powerful grip on the human imagination – and it was another science fiction author, Isaac Asimov, who would perhaps do most to popularize the idea that they could become a matter of fact as well as fiction. In his 1942 short story 'Runaround', Asimov coined a new word – 'robotics' – to give a name for the study of robots.[13]

'Runaround' also coined what Asimov called the Three Laws of Robotics as a central plot device: that robots must not injure a human being, or allow them to come to harm through inaction; that they must obey orders from humans, except where these would conflict with the first law; and that they should protect their own existence, so long as this did not conflict with the first two laws.

Asimov's three laws both helped give birth to an entire genre of speculative fiction and fuelled interest in what would become a major field of scientific research, culminating in 2000 in the commercial release of one of the first truly humanoid, walking robots by the Japanese company Honda. Delightfully – and apparently entirely coincidentally – this robot's name was ASIMO, an acronym for 'Advanced Step in Innovative Mobility'. Sadly, the American company US Robotics – although it was named specifically in honour of Asimov – doesn't actually make any robots at all.

# NETYMOLOGY

Between them, 'robots' and 'robotics' have spread throughout the world, sometimes in unlikely contexts. If you're driving along a road in South Africa and see a painted sign warning of a 'ROBOT' ahead, don't expect a humanoid machine to be waiting round the corner. The word is today commonly used as an abbreviation for the phrase 'robot policemen', which is what traffic lights were first called when they were introduced to the country.

# 12.

# Cyber-everything

The prefix 'cyber-' is found everywhere today as an all-purpose term for digital and internet activities. Teenagers persecuted on Facebook are suffering from 'cyber-bullying'; other teenagers banned from accessing the internet by their parents, perhaps because they've been bullying classmates on Facebook, have been 'cyber-grounded'. Those engaging in advanced forms of flirtation via their computer screens, or worse, may say they're having 'cyber-sex'; those wanting to be portentous about all things digital may talk about the 'cyberverse'.

The variations are almost endless, and can extend deeply into life beyond the screen: from pervasive concerns over 'cyber-security' and 'cyber-stalking' to real world 'cyber-parks' in which a large number of technology companies are concentrated.

All of these are terms of the last few decades. 'Cyber' itself, however, has an impressively long etymological pedigree. The prefix featured for the first time in the English language as part of the word 'cybernetics' — coined in

1948 by the American mathematician Norbert Wiener in his book of the same name, which explored the properties of self-regulating information systems.

Wiener borrowed 'cybernetics' directly from ancient Greek, in which a *kybernetes* was both the person who steered a ship and the rudder of that ship. Wiener was using the word in a metaphorical sense, originally deployed by no less an authority than Plato to describe the governing or 'steering' of people themselves.

Cybernetics, then, became a way of speaking about the possibilities of self-steering information systems – and by the 1970s it had become a familiar enough term among specialists for a range of supercomputers to be released by manufacturers CDC under the brand 'Cyber'.

It wasn't until its adoption by a young science fiction author in 1982, however, that 'cyber-' truly broke into mainstream culture. The author was the American-Canadian William Gibson, who in his story 'Burning Chrome' coined a tantalizing term for new realms of human-technological interaction: 'cyberspace'.[14]

It was a word that truly took off two years later thanks to its central role in Gibson's debut 1984 novel *Neuromancer*[15] – a book which successfully predicted, and indeed helped shape, whole swathes of what would become the popular cultural movement known as 'cyberpunk'.

Confusingly enough, the term 'cyberpunk' itself predates both *Neuromancer* and 'Burning Chrome', originating instead in a 1980 short story of that name by author Bruce

Bethke about a gang of teenage computer hackers. Bethke's story was not actually published until November 1983, just in time for its title to be indelibly taken up as a description of the aesthetic of *Neuromancer* and its countless imitators and innovators.

These range from films like 1999's *The Matrix* (whose title is itself a direct borrowing from the name of the computer network in *Neuromancer*) to the novels of authors like Neal Stephenson and Bruce Sterling, not to mention video games, Japanese anime strips and films, music videos and albums, and even architectural and fashion styles. Simply Google 'cyberpunk fashion' and its more recent relative, 'cybergoth', for more leather, metal and plastic than you can shake several modems at.

# 13.
# Three-letter words

Typing and texting have done wonders for the art of abbreviation, as even the *Oxford English Dictionary* has been forced to acknowledge. In March 2011 the English language's dictionary of record selected for publication 'a number of noteworthy initialisms' of which two should bring particular delight to digital eyes: OMG ('Oh my God', first recorded, amazingly enough, in a 1917 letter from a British admiral to Winston Churchill) and LOL ('laughing out loud', first recorded in the mid-1980s on a Canadian online Bulletin Board System, courtesy of one Wayne Pearson), six letters which between them summarize much of the emotional flavour of current casual typed communications.[16]

An initialism like LOL is distinct from an acronym like Scuba (which stands for 'self-contained underwater breathing apparatus') because it creates not a new word as such, but a collection of capital letters that function as a single time-saving unit.

Soldiers writing home in the Second World War pop-

ularized such pre-digital initialisms as SWALK (Sealed
With A Loving Kiss), MALAYA (My Ardent Lips Await
Your Arrival) and the rather more forward BURMA (Be
Upstairs Ready My Angel). Text messages, emails and
online chat have, however, ushered in a whole new level
of reference and self-reference of which it's only possible
to scratch the surface here, but which is characterized
above all by the steady accretion of letters.

Consider an early enhanced version of LOL, for when
things get really amusing: ROFL ('rolling on the floor
laughing'). In an inexorable process of escalation, this soon
gained its own superior variant, ROFLMAO ('rolling on
the floor laughing my ass off') – onto which those wanting
to indicate bored amusement with the whole business of
online initialisms sometimes graft three further letters
to make ROFLMAOBBQ ('rolling on the floor laughing
my ass off barbecue'). Its ridiculousness is a large part
of the point, alongside the need to differentiate one's
own feelings from mere everyday LOLlers; see also
ROFLMAOASOFBS, for 'rolling on the floor laughing
my ass off at someone's Facebook status', and countless
similar variations.

Originally typed in order to convey emotional states,
LOL and its countless relatives have today become sufficiently accepted to start counting as conversational words
in their own right, thus blurring their way ever closer
towards the status of full acronyms. 'Lol' is usually spoken
to rhyme with 'doll', accompanied by some kind of ironic

facial expression. Similarly, the explanation that you're doing something 'for the LOLs' or 'for the lulz' has become a global shorthand for doing something stupid for its own sake.

There are also rich international pickings to be had here. French has its own variant, MDR, standing for *mort de rire* or 'dying of laughter'; while Spanish tends to go for 'jajaja' (pronounced close to 'hahaha' in English) or 'jejeje' (in English, 'hehehe'). In Thailand, typing '555' also reads approximately out loud as 'hahaha', the Thai word for five being *ha*; while Swedes go for 'asg', an abbreviation of the Swedish word *asgarv*, or 'roars of laughter'.

In the Philippines, meanwhile, an entire youth culture has drawn its name from these abbreviations: the 'jejemons', a term derived from a combination of the Spanish version of LOL, *jejeje*, with the abbreviation of the word monster into 'mon'.

*Jejemons* epitomize a particular contemporary cultural strand: the use of sociolects (language varieties restricted to particular social groups) based on text messages which infuse English and international terms with local words. 'iMiszqcKyuH' means, for example, 'I miss you', via an alchemical mix of phonetic English, deliberately reordered and mixed case letters, and a general desire to be as incomprehensible to outsiders as possible.

# 14.

# Everyone's an avatar

One of the most mystically charged of modern tech coinages is the word 'avatar' – a term popularized by James Cameron's record-breaking 2010 film of the same name, but in use for several decades before that in the realm of computer gaming and virtual worlds.

An avatar is a player's digital incarnation within a virtual world, and has its origins in the Sanskrit term *avatara*, meaning 'descent', used to describe the descent of a god from the heavens into some kind of earthly form. In Hindu mythology, gods frequently take on earthly incarnations, ranging from animals to warriors and lovers, and the word has been used in English since the late eighteenth century in this mythological context.

This was the sense of the word 'avatar' when used in the title of science fiction author Poul Anderson's 1978 novel *The Avatar*, referring to an alien being taking on a new form. The word, however, had entered the genre fiction consciousness – and in 1992 the author Neal Stephenson published his novel *Snow Crash*, which for the first time popularized it in a technological rather than mystical sense.

*Snow Crash* is a set in a near future in which citizens explore an alternate digital reality via user-controlled avatars, a use of the word fusing together notions of computation and incarnation – and giving flesh to the growing realization that computers were becoming as much a portal to a species of shared reality as they were mere tools.

As the acknowledgements to *Snow Crash* note, the term avatar had actually been in use for some time – unbeknownst to Stephenson himself – thanks to the world's first graphical online role-playing game, 1985's 'Habitat'. Stephenson did, however, coin another influential term to describe *Snow Crash*'s three-dimensional computer-generated world: the 'Metaverse', using the Greek prefix *meta-* (meaning 'beyond') to replace the prefix *uni-* (meaning 'one') in the word 'universe'.[17]

It takes some time for reality to catch up with fiction, and the first 'real' computer avatars were little more than crude, two-dimensional images representing users in chat rooms and discussion forums. Since then, although we are still some way from living up to James Cameron's fantasies, human embodiment within digital environments has grown increasingly in sophistication, thanks most recently to the power of devices like Microsoft's Kinect to translate the movement of a user's body in real time into an onscreen presence.

Indeed, the word 'avatar' is now sufficiently generic that

some manufacturers have taken to inventing their own terms for proprietary virtual incarnations. Witness Nintendo's introduction of cartoonish three-dimensional computer characters for its Wii console called – inexorably – 'Miis', a word with all the etymological resonance of a full stop.

# 15.
# On memes

In his 1976 book *The Selfish Gene*, evolutionary biologist Richard Dawkins gifted the world more than even he might have bargained for when he declared his intention to coin 'a noun that conveys the idea of a unit of cultural transmission, or a unit of imitation. *Mimeme* comes from a suitable Greek root, but I want a monosyllable that sounds a bit like "gene". I hope my classicist friends will forgive me if I abbreviate *mimeme* to *meme*.'[18]

Dawkins may or may not have been forgiven by classicists, but one thing he couldn't have predicted would be that his new word (the cousin of a similar word coined by the German biologist Richard Semon in 1904, also from the Greek: *mneme*) would come to be one of the most useful and defining terms of twenty-first-century internet culture. For, in a digital age, 'memes' are a uniquely contemporary class of object – a kind of pop cultural flotsam, spreading via all those they persuade to laugh out loud along their path.

There are tens of thousands of memes online,

embodying near-unimaginable quantities of ingenious timewasting. From dancing babies to flying felines shaped like pop tarts, they sweep from satire (the gap between politicians releasing campaign posters and the net adapting them into parodies is now numbered in minutes rather than hours) to pop-cultural gags with more layers than a set of Russian dolls. Read the 3,000-word Wikipedia article on 'Rickrolling' for an example of how a meme can eat itself several times over within the space of five years.

Memes have not only fans by the million, but their chroniclers too. The definitive point of reference for all matters meme-like is perhaps knowyourmeme.com, which boasts of its dedication to 'documenting Internet phenomena'. With over 1,200 confirmed meme entries, it's a repository of human inventiveness and strangeness bordering on the pathological; not to mention a treasure trove of wonderful would-be words. A 'kitler' is, for example, a 'cat that looks like Hitler', and some of the photos are uncanny.

There's much argument over the first digital meme – although one leading contender is the emoticon for a smiley face, :-). There's little debate, though, over the most influential of all memes: the infamous 'lolcat'.

Lolcats are the near-numberless offspring of a venerable class of object: the 'image macro', in which text is superimposed on a photograph. Born around 2006 on the notoriously obscene and inventive imageboard website 4chan.org, 'laugh-out-loud-cats' – as nobody would ever dream of spelling them out – pair cute animal images with comically misspelt captions. This may seem distinctly limited grounds for amusement. But type 'lolcat' into Google and you'll turn up not only six-million-plus pictures, but initiatives ranging from scholarly studies to the frankly bewildering 'lolcat bible translation project'.

Like the capacity of human society itself to act, in Dawkins's terms, as a kind of gene pool for thoughts and belief, we are today the conscious agents of our own delight and distraction in a way never previously possible. And with this comes a new sense of what it means to be part of a global human collective, celebrating and rethinking its own nature every day – comic cats and dancing babies included.

# 16.
# Hacking through the net

The word 'hack' is an ancient one. So long as you're referring to chopping something to pieces, it's been essentially
unchanged in English for eight hundred years, and is
thought to be traceable back many thousands of years before
that to an ancient pan-European term for a tooth or hook.

In computing terms, it's a far younger idea — but one
with a surprisingly similar origin. Because digital 'hacking'
started out meaning broadly what it always had: chopping something down to size and then rearranging it at
will, quite possibly with mischievous intentions.

From the early 1960s, talk of 'hacks' (used as a noun)
meaning clever tricks or pranks was reported at the
Massachusetts Institute of Technology. Early computers
had extremely limited resources for running programs, as
well as vulnerabilities that could be exploited by those in
the know. Performing clever 'hacks' was thus the sign of
a true expert and innovator: someone able not only to
understand a system completely, but to think laterally
around its limitations.

By the late 1970s, pioneers of home computing like Apple founder Steve Jobs were starting their careers in parents' garages, dismantling and rebuilding electronics – 'hacking' in the most literal of senses, and expanding the associations of the word in the process towards the dawning realm of home computing.

By the time of the world wide web's arrival in the 1990s, hacking's darker connotations were coming increasingly to the fore in its usage – thanks partly to scare stories in the media, partly to fiction about those seeking to undermine world order from within, and partly to real malice among those determined to exploit the world's rapidly increasing number of inexperienced computer users.

To those in the know, determining hacking's precise connotations is a much-debated controversy, while 'hacker culture' itself represents a series of traditions encompassing many of the ideas in this book (most iconically embodied in a document known as the 'Jargon File': see the bibliography for further details). Yet it's also a pursuit that has always come in different 'flavours' – and these have over time been formed into a semi-official classification, based on the old-fashioned western movie trope of baddies and goodies wearing different coloured hats.

As you might expect, 'black hat' hackers are the worst of all digital baddies: someone breaking into computer systems for no reason beyond malice or personal gain. Their counterpart is a 'white hat' hacker, whose ethical approach to hacking involves testing and aiming to improve computer security systems, usually in an official capacity.

Between these two lies the realm of the 'grey hats', who may unofficially take on the kind of work performed by white hats – even if they haven't been invited to do so. There can also be 'blue hat' hackers – employed by security consulting firms to test systems – and, leaving hats behind for a moment, 'elite' hackers at the very top of the global pecking order, with callow 'noobs' at its base (see entry number 41 for more detailed commentary on this delightful term).

Perhaps most pleasing of all, at least linguistically, is a term for those using hacking to send out a political message which has come into its own over the last five years: 'hacktivism'. In a world where shutting down the website of a credit-card company can be a potent form of protest, this portmanteau of 'hacking' and 'activism' is perhaps the perfect word for our times.

# 17.
# Do you grok it?

One of the more unusual gifts from the realm of science fiction to mainstream nerd vocabulary is the notion of 'grokking'. To 'grok' something is to understand it so completely that, in the words of its coiner, Robert A. Heinlein, 'the observer becomes a part of the process being observed'.

Heinlein first deployed the word in his 1961 novel *Stranger in a Strange Land*, and set out to create an alien-sounding original term – befitting its Martian origins – describing a state of extreme and intimate empathy itself alien to mere 'Earthlings'. While its literal meaning is 'to drink', grokking has, Heinlein explains, the metaphorical sense of 'to drink in all available aspects of reality' – something quite impossible for the unenlightened human mind.[19]

Heinlein, however, reckoned without the rise of a computer programming culture in which claims of complete – indeed, almost mystical – understanding were to become a vital part of the coding ethos.

Simply to understand how to use a particular programming language is one thing. But by the 1980s, the claim that you 'grokked' how to do something on a computer had become a geek declaration of supreme expertise – of a profound grasp not only of a programming language, but of the very world-view and philosophy it embodied.

This notion of computing as something with which one can have a deeply empathetic relationship has become a staple of the culture of computer science, where the most advanced acts of programming are essentially a process of world-building. It's not for nothing that one common term in use to describe high-level program designers is 'software architects': people putting together virtual entities as seamless and sophisticated as entire buildings.

It's an attitude reflected in many of the terms used to describe computer programming, where notions of elegance and integrity can be almost as important as mere functionality. Consider the distinction between an 'elegant' solution to a problem and a 'kluge' or 'kludge'. The latter word is taken from the German *klug*, meaning 'clever', but whose use in programming refers to a non-optimal solution, involving cleverly working around a problem rather than dealing with it directly. It means, in effect, achieving the right result for the wrong reasons. If you're forced to come up with a kludge, you probably haven't fully grokked the problem.

Similarly, terms of aesthetic revulsion abound in the informal language of computer expertise. Bad code might make you 'barf' in disgust – or 'puke' or 'vom' – while you could be unlucky enough to encounter the 'crawling horror' of outdated, unworkable software kept alive by the dark forces of those who know nothing about computing (the latter term being a borrowing from American horror writer H. P. Lovecraft).

In each case, the level of emotional involvement with the world of the machine is what's most evident, together with a visceral response to its pollution: something even Robert A. Heinlein's Martians might have admired.

# 18.

# Sock puppets and astroturf

'Online', the old joke goes, 'nobody knows you're a dog' – and deception is certainly a rich digital field, complete with its own vocabulary of feints, fallacies and carefully fostered illusions.

Once upon a time, sock puppets were exactly what the phrase suggests: puppets made out of old socks, worn on the hands for family entertainment. Today, however, they've developed a dirtier digital second life, as a term for false online identities wielded by a cynical puppeteer in order to make themselves and their ideas look better, or make their opponents look worse.

Perhaps the most (in)famous sock puppet of recent times was a young Syrian blogger going by the name of Amina Arraf. Arraf's blog – A Gay Girl in Damascus – first appeared in February 2011, and appeared to tell the story of a young lesbian woman living amid a climate of growing political protest and state crackdowns. Arraf gained a cult following with her tales of illicit love affairs and pursuit by the state security forces, a narrative culminating in

June 2011 when a post appeared under the name of her cousin, claiming that Arraf herself had been taken captive by armed men on a street.

It was a shocking tale that grabbed international media interest – until, under investigation, details of the elusive Arraf's life began to unravel. Following extensive digital detective work, a forty-year-old man called Tom MacMaster, studying for a masters at Edinburgh University, confessed that Amina Arraf was his fictional creation. A heterosexual American, MacMaster had, he claimed, 'got caught up in the moment'.[20]

Although it caused outrage, MacMaster's puppetry represents one of the more innocent variations of the art, in that its intentions were as much creative as deceiving. Far more cynically conceived are the legions of false identities that haunt review sites and discussion forums, puffing products and services; or that may be deployed by governments and intelligence agencies to flush out dissidents by cultivating the illusion of safe, sympathetic listening.

There's even a label for this process of deception when taken to a mass-manufactured scale: 'astroturfing'. Invoking the artificial grass on which sports are often played, it neatly conjures the image of unseen puppeteers creating the false impression of a grass-roots movement online through multiple false identities. Whether they're cynical exercises in publicity, marketing or politics, astroturfing campaigns can be highly effective, laying down a false layer of seemingly 'natural' information that to many users might be indistinguishable from a real groundswell of comment (the term itself reportedly dates back to a 1985 usage by Senator Lloyd Bentsen, describing suspiciously co-ordinated postal campaigns by 'ordinary' citizens).

We are, after all, never quite ourselves when acting and reacting onscreen. Rather, we're performers, taking our turn at trying out different roles and tools – and wondering what socks suit us best. This makes puppetry itself (from the Latin, *pupa*, 'a little girl or doll') both a fine metaphor and caution for many online acts, with unseen fingers pulling the strings of their diminutive onscreen agents – the entire process lifeless without people to gift it action and purpose.

# 19.
# Bluetooth

If you're looking for an obscure icon for the digital age, the tenth-century Danish king Harald Gormsson fits the bill better than most. His connection to the cutting edge becomes considerably clearer once you add his nickname *blåtand*, or 'bluetooth' (earned thanks, it's rumoured, either to an unpleasant gum disease, a dark complexion, a fondness for eating blueberries, or all three). For 'bluetooth' was the name given in 1994 by Swedish company Ericsson to a new wireless protocol for exchanging data over short distances – one that would, they hoped, live up to the standards set by its namesake.

In the tenth century, Harald 'Bluetooth' Gormsson's great achievement was the unification of warring Danish tribes under his rule. Similarly, Ericsson's symbolic hope for its Bluetooth was to unify the 'warring' range of protocols for wireless communication into a single, universal standard – something that Bluetooth has since gone a long way to achieving, being found today in over seven billion different devices around the world.

# Bluetooth

Bluetooth's logo plays off this tradition, combining the Scandinavian runes Hagall and Bjarkan – King Harald's initials – to make a single 'bind rune'. It also makes a pleasant change from the distinctly conservative naming policy usually found among Scandinavian tech giants, which tend to be dubbed either after their founders (as in the case of Ericsson, named after the nineteenth-century Swedish inventor Lars Magnus Ericsson) or the place of their founding (as in the case of Nokia, named after the small town in south-west Finland).

As ever with language, widespread adoption breeds variation – and, in the case of technology, the rapid import of concepts and variants from elsewhere in the field. Tellingly, verbal proliferation in the case of Bluetooth has centred on cracking, hacking and generally pushing the system to breaking point: perhaps the most fertile area for innovation in all modes of communication.

Within the pleasing pool of terms that Bluetooth's success has spawned, 'bluesnarfing' is one of the most widely used – describing the unauthorized 'snarfing' (a slang term for stealing or grabbing, first recorded in the 1980s) of information being transmitted across a Bluetooth connection.

Less serious than bluesnarfing – although still a cause for concern – is 'bluejacking', which means transmitting unauthorized information to a Bluetooth device in order, effectively, to send a spam message to an unwitting recipient. Finally, and potentially worst of all, comes 'bluebugging' – which means attempting to manipulate someone's phone or digital device via Bluetooth in order to take it over completely.

It may sound amusing, but a BlueBug assault is not to be sniffed at, threatening complete loss of control of one's mobile phone within just a few seconds – accompanied perhaps by a different kind of blue language altogether.

# 20.

# The Cupertino effect

For language lovers, it can be thrilling to find a name for something you've often observed but never been able to label — and the so-called 'Cupertino effect' is a prime contemporary example. It describes one of the most unlikely gifts of the age of word processing: the way in which an over-zealous automatic spellchecker can 'correct' your attempt at typing one word into something similar, yet crucially different.

The term itself is taken from the northern Californian city of Cupertino — home to no less a company than Apple — thanks to the omission of the word 'cooperation' (when spelt without a hyphen dividing co- and -operation) from the dictionaries of some early spellchecker programs.

This meant that, whenever 'cooperation' appeared in a text, it would be flagged as incorrect. More significantly, for those users who had set their word processors automatically to adjust their typing into 'correct' spellings, the word would be changed to 'Cupertino': the name of the city no doubt featuring in the software's dictionary thanks to its illustrious corporate residents.

Search for the word Cupertino online, even today, and you'll soon unearth such delights as 'reinforcing bilateral and multinational Cupertino' from official documents created in the early 1990s; or attempts to develop 'quality education by encouraging Cupertino between Member States' – this last one courtesy of a European Parliament working paper on languages.[21]

Needless to say, the wired world also abounds with broader examples of the genre. One of my favourites has always been the fondness of an early word processor for turning 'Freud' into 'fraud'. Even today, technology is no guard against error if left unchecked. Witness this correction from the *New York Times* in April 2012, as highlighted under the title 'tasty cupertinos' by the Language Log blog: 'An earlier version of this article incorrectly referred to the Ethiopian dish doro wot as door wot. Additionally, the article referred incorrectly to awaze tibs as aware ties.'

Computer spellchecking hasn't stood still, and considerable research has been poured into making it smarter and more sensitive to context. Fortunately for Cupertino connoisseurs, however, the modern proliferation of devices on which people type has brought with it a whole new spectrum of spellchecked error. Some of the most pleasing are collected on sites like http://cupertinoeffect.tumblr.com, which bills itself as 'The running log of what I type into my iPhone and what the Apple Computer Corporation of Cupertino, CA knows I want to say'.

Favourites here include 'Yankee fandom' converted into 'tanker random', 'gangbang' to 'handheld', 'yogi' to 'tofu', and 'yessir' into 'yessiree'. Humans being what they are, it can be difficult not to read such corrections as an oblique creative force in its own right — and, sometimes, an ironical commentary on what you're trying to say. Until it learns the word, for example, some phones can transform the company Facebook into 'ravenous' — surely a subliminal hint against the consuming tendencies of social media.

Cupertinos can also be a serious business. Consider the case in 2012 of a school in the state of Georgia which was evacuated after a text message reading 'gunman be at west hall today' was reported to police. The message, it transpired, had been supposed to read 'gonna be at west hall today', but had been automatically 'corrected' without the sender noticing.[22]

# 21.
# The Scunthorpe problem

Much like automatic spelling corrections, the idea of filtering digital content in order automatically to remove sources of offence is almost as old as computing itself. Long before the world wide web came along, internet forums and bulletin board systems provided a seeding ground for everything from pornographic images to extremely frank exchanges of views. Hence the need for some protection for those of delicate sensibilities – and hence the birth of the marvellously named 'Scunthorpe problem', whereby entirely innocent words and phrases fall victim to machine filth-filters thanks to unfortunate sequences of letters within them.

In Scunthorpe's case, it's the second to fifth letters that create the problem – a phenomenon first spotted in 1996, when America Online's internet service temporarily prevented anyone living in Scunthorpe (an English town found in the north of the county of Lincolnshire) from creating user accounts.

Scunthorpe itself innocently draws its name from the

# The Scunthorpe problem

Old Norse word *Escumetorp*, meaning 'The homestead belonging to Skuma', a name coined during Danish rule of the north-east of England during the ninth to eleventh centuries. This venerable history did not, however, stop Google from echoing AOL's anti-Scunthorpe stance, with its SafeSearch filter restricting search results for businesses with Scunthorpe in their names as recently as 2004.[23]

It's not just Scunthorpe that has suffered from eponymous problems. Pity the residents of, for example, Penistone in South Yorkshire; or indeed any innocent member of the public with a name like Cockburn, whose first four letters may still be deemed too obscene for creating an email account by some providers.

More elaborate than mere blocking, however, is the automated substitution that can sometimes take place online when words or phrases are identified as 'offensive'.

Consider the fate of American sprinter Tyson Gay, whose name was automatically converted to 'Tyson homosexual' throughout an article appearing on the website of the American Family Association.[24]

As the last example suggests, perhaps the most interesting thing about automated prudishness is what it reveals about the intentions of the people designing a system of censorship in the first place – and when it reveals a filtering process that might otherwise have remained invisible.

This last phenomenon is sometimes referred to as 'the clbuttic effect', named in honour of the mangling of the word 'classic' by over-zealous obscenity filters. For the mistake to occur, the letters 'ass' are transformed into the (presumably less offensive) word 'butt', leaving you with a text full of 'clbuttics' – not to mention the potential for 'mbuttive' (massive) quantities, someone 'pbutting' (passing) a football, and so on.

Google away, and you'll soon find almost 5,000 instances of the 'Buttociated Press' online (instead of the Associated Press) and almost 200 mentions of a unique gift of the digital age's human-machine collaborations, the 'personal buttistant' – a far more intriguing term than any mere PA.

# 22.

# The coming of the geeks

The meaning of all words shifts over time. Sometimes, terms that were once forceful expressions of disapproval – like 'bad' and 'naughty' – become watered down with milder connotations. Terms of praise or exaltation – such as 'awesome' or 'incredible' – may also lose their original force. Words may even reverse their meaning, such as the adoption of 'wicked' for praise rather than blame.

Then comes a more explicitly cultural phenomenon: the gradual upward slide of an outsider term into mainstream culture. Popular music saw such transitions over the twentieth century, with forms like jazz, funk and blues all achieving mainstream acceptability decades after their origins at the margins of society. As we begin our travels through the twenty-first century, a similar transformation is taking place in the technological realm – for we are witnesses to the rise and rise of the geek.

'Geek' arrived in English from Low German, in which a *geck* denoted a crazy person or fool, and it was this sense that provided its first English meaning. In the subculture of nineteenth-century carnivals, geeks were outsiders even among the outsiders, known for their bizarre and often disgusting acts. In early American travelling circuses, the 'geek show' traditionally involved a performer biting off and then eating the heads of live chickens.

Come the 1980s, carnivals of this type were (thankfully) a thing of the distant past, but the term 'geek' found itself revived in America as a label for socially awkward children obsessed with new technological devices. The pejorative sense remained, for a while. But, as a generation of tech-savvy youngsters began to provide the first generation of internet millionaires, and then billionaires, the unthinkable happened: geeks began to become cool.

It's difficult to put a precise date on it – but the modern world has embraced both the word and the notion of geek culture (not to mention its conveniently rhyming fashion associate, 'geek chic').

As much as anything, the transition is an index of technology's infiltration of almost every aspect of popular culture. Apple's first iPod appeared in 2001, and rapidly became an icon. Today, the notion of buying new music via anything other than digital download feels quaint, or at least a willed exercise in nostalgia. Video games – once the very embodiment of a marginal, antisocial cultural force – have become both mainstream and fashionable.

Interestingly, the supremacy of 'geek' has come partly at the expense of another word used to describe the technologically astute but socially inept: 'nerd'. This term is considerably younger than 'geek', being first recorded in 1950 in the Dr Seuss story 'If I Ran the Zoo', where it denoted a strange creature (from the land of Ka-Troo). On a quite possibly unrelated note, it first seems to have been recorded in the slang sense of an unfashionable outsider the following year.

By the late 1960s, 'nerd' had begun to achieve pop-cultural prominence, boosted still further by its usage in the 1970s sitcom *Happy Days* and later — perhaps its linguistic apotheosis — in the title of the 1984 satirical film *Revenge of the Nerds*, cementing its status as the term of choice for all smart-but-inept citizens.

Today, perhaps because of this firm anchoring in 1980s culture, nerd just sounds a little too quaint and insufficiently technological to do justice to twenty-first-century culture. The nerds may have had their revenge but, as headline writers have been punning at least since the 1990s, it's geeks that are set to inherit the earth.

# 23.
# Beware of the troll

For those venturing into any form of online discussion, 'do not feed the trolls' (sometimes simply abbreviated to DNFTT) remains one of the oldest and best pieces of advice around.

In the pre-web days of the 1980s internet, 'trolling' originally meant logging into online forums and posting deliberately naive or provocative questions, in order to bait people who didn't realize they were being 'trolled' into long, frustrating discussions.

Over time, online trolling came to refer to ever-worse behaviour, and the use of 'trolls' as a noun arose to describe anyone who maliciously attempted to sow digital dissent through their words. Hence the injunction not to 'feed the trolls' – that is, not to respond to malicious provocation.

This latter sense of the word aligns it closely with its mythological Scandinavian namesake. The Old Norse term *troll* referred to a wide class of demons, monsters and malefactors, linked to an entire taxonomy of magic-wielding *troll-women* and others able to bewitch the hapless. There's even a modern Swedish word, *trolla*, meaning 'to charm or cast a spell upon'.

Despite these common resonances, however, the online term didn't begin as an imitation of Nordic fiends. Rather, 'trolling' can be traced back to the Old French verb *troller*, meaning to wander around aimlessly: a term that by around 1600 was being used in English to describe fishing from the back of a moving boat, and thus ranging freely across a body of water in the hope of attracting a bite.

By the late 1960s, this kind of 'trolling' had extended its metaphorical reach to people wandering around in the hope of a sexual encounter. And it was this sense of a seemingly casual but in fact strategically targeted entice-ment that helped consolidate the use of trolling as a description of dubious online enticements.

Since then, the birth of the troll as an internet bogeyman has brought its other associations into play, and cartoon illustrations of vile 'internet trolls' abound online. Inex-orably, too, trolling's intensification as a phenomenon has included a migration from the strictly amateur domain of mischief-makers to what's sometimes known today as 'professional trolling'.

A professional troll is someone who posts material on-line – from blogging or journalism to speeches and inter-views – that's expressly designed to provoke controversy and comment, and to bait an audience into discussion.

It's a habit that connects some modern trolling to another fishing-related metaphor on the darker side of the internet, 'link-baiting': using words, topics, arguments and ideas on your website expressly in order to attract attention and to try to get other people to link to you. This can range from the crudest tactics ('sex!') to more sophisticated attention-seeking arts (monitoring and then dropping the name of anything that happens to be trending on social media at a particular moment).

All this has also helped push the idea of trolling beyond strictly online activities, allowing the word increasingly to be used as a description of any form of cynical attention-baiting – from self-publicists on TV news deliberately courting controversy to authors making outrageous claims. Given the internet's rise as the engine driving all other news and attention cycles, it perhaps shouldn't be surprising that teaching the world to troll may prove one of its most enduring legacies.

# 24.
# Bitten by bugs

If you're looking for an insight into the mindset of many modern computer programmers, you could do worse than start with how they name their nemeses: bugs.

The word 'bug' itself, when used to describe a fault within machinery, is far older than modern computing. As early as 1878, the American inventor Thomas Edison was writing to an associate about 'bugs' occurring in the process of invention. 'The first step is an intuition, and comes with a burst, then difficulties arise – this thing gives out and [it is] then that "Bugs" – as such little faults and difficulties are called – show themselves,' he explained in a letter to Tivadar Puskás, his agent in Paris, describing his continuing efforts to develop a practical electric light.[25]

Edison's familiar use of the term suggests something already known to engineers, and it continued to be used in a machine context through the first half of the twentieth century (unlike the younger term 'glitch', which entered English – from the Yiddish word *glitsh*, 'a slip' –

in the 1960s, as a word used by electronic engineers and popularized by the US space programme).

'Bug' got its first outing as a computer term pretty much as soon as electronic computers arrived, and was spread via a story told by the computer science pioneer Grace Hopper, who in 1947 encountered a log book by a fellow member of the team working on the Harvard University Mark II Aiken Relay Calculator. The entry recorded an error being fixed by the removal of a dead moth from the machine's workings – a moth physically affixed to the log book next to the caption 'first actual case of bug being found'.[26]

Perhaps the world's most literal case of successful debugging, Hopper's tale suggests an engineering culture that had already begun to breed variations, subsets, and playful reinterpretations. Today, that taxonomy has reached a scale appropriate to the many varieties of electronic error: and, as you might expect, its margins boast some fine oddities.

One notable such term is the 'Heisenbug' – named after Werner Heisenberg's eponymous uncertainty principle because it changes its behaviour the moment a programmer attempts to isolate it, effectively vanishing the moment it is observed.

Continuing the theme, a 'Bohr bug' – named after Danish physicist Niels Bohr – behaves in the opposite way to a Heisenbug, being easy to reproduce under defined conditions (by analogy with Niels Bohr's predictable, deterministic model of an atom).

## Bitten by bugs

Then, named after mathematician Benoît Mandelbrot, come 'mandelbugs' – programming errors so complex that they appear chaotic, like the infinite layers of fractal complexity found in a Mandelbrot set.

More mysterious still are 'Schroedingbugs' – named after Erwin Schrödinger's famous thought-experiment involving a cat that manages to be alive and dead at the same time. Similarly, a Schroedingbug can lurk for years within an apparently perfect piece of computer programming, until someone notices an esoteric error that ought to stop it from working – at which point the program does, indeed, stop working.[27]

# 25.

# Bits, bytes and other delights

Almost all electronic data is ultimately composed of ones and zeros. In one sense, this is the basis of modern computing: the encoding of all the world's information in a binary format (from the Latin *binarius*, 'consisting of two things'). No matter how complex the digital tools and services we use, the two basic states of electronic charge underpin them. Something is either switched on, and thus has a positive electrical charge (representing a 'one' in the binary counting system); or it is switched off, and thus holds no charge at all (a binary zero).

At the opposite end of the scale to binary is 'analogue' – a term derived from the Greek words *ana* ('up to') and *logos* ('word', 'account' or 'ratio'). As the twinned English term 'analogy' suggests, the analogue representation of something involves not an impassive code of ones and zeros, but a record that is physically analogous to the nature of its original. Undulating sound waves carved into the vinyl surface of a record, for example, or the intentionally imperfect business of sketching a scene by hand onto paper.

## Bits, bytes and other delights

Storing information in an analogue format means creating a new, unique object for every subject you wish to represent: a new sketch, recording, picture, and so on. Storing information digitally is quite different because the ones and zeros of digital charge are, on a fundamental level, compatible with each other. Given the correct hardware and software, digital information of all types can be reproduced, altered, exchanged, stored and merged almost infinitely.

All of which involves increasingly large collections of ones and zeros, the most fundamental unit of which was christened a 'bit' – short for 'binary digit' – in the late 1940s by the American statistician John Tukey, while he was working at Bell Labs (Tukey was also the first person to use the word 'software' in print, among his other claims to linguistic fame).

Given that bits can only have two values, they're so small that even early electronic computers tended to process them several at a time. Thus was born the gently punning term for a collection of bits, a 'byte'. It's a term whose precise origins are uncertain, but it is sometimes attributed to the computer scientist Werner Buchholz in 1956, and his desire to coin a new word that was closely connected to 'bit' but couldn't be confused with the existing word 'bite' when written down.

The number of bits in a byte initially varied depending on the computer manufacturer, but soon settled on a standard eight, allowing for 256 different potential values to be encoded within each byte. From this soon emerged a whimsical term for half of a standard byte, or four bits of information – the 'nibble', a unit not officially used in computing, but affectionately nodded towards within many computer science courses.

With equal whimsy, a collection of 4 bytes is sometimes referred to as a 'word' by programmers; while two bits is said to represent a 'shave and a haircut', as per the 120-year-old music hall riff (an etymology that really has to be hummed in order to be appreciated fully).

# 26.

# Twinks, twinked and twinking

'You've twinked your character'; 'I'm going to login with my twink' – standard phrases in the realm of online games, where specialized and unlikely-sounding vocabulary abounds.

'Twink' – which can function variously as noun, verb and adjective – belongs to an etymologically rich field: words describing gaming power and patronage. It occurs when an experienced player uses the wealth of in-game resources earned by a powerful character to bestow exceptionally good equipment upon one of their lower-level, secondary characters. This points in turn towards a decidedly non-technical origin for the term: in the identical gay slang for an attractive, effeminate young man in receipt of the patronage of a generous older 'sugar daddy'.

Twinks, in this sense, date back to at least the early 1960s – and may, the *OED* notes, be related to the more innocent idea of something attractive 'twinkling' (a word present in English since the fourteenth century, and derived from the Old English *twinclian*, 'to wink or blink').

A popular alternative etymological theory is that both the gay slang and the gaming slang originate from the American snack cake brand Twinkie, manufactured under that name since the 1930s, whose soft creamy filling and quasi-phallic shape certainly make for some interesting associations. It may, however, simply be the case that Twinkies' existence reinforced rather than originated the slang terms.

Much as gay slang offers a comprehensive taxonomy of types and tastes, gamer slang also abounds with labels for different playing styles, one of the more colourful of which is the 'Munchkin' type: describing a player who seems determined to behave ferociously competitively within a game world or setting intended to be non-competitive.

The term 'Munchkin' itself is a loan from L. Frank Baum's 1900 children's classic *The Wonderful Wizard of Oz*, and seems first to have been used within gaming discussions during the earliest days of the internet as a dismissive term among older, more experienced, players for those Munchkin-like new players with an intemperate devotion to progress at all costs.

Munchkins may well be found twinking their characters in order to get ahead, or indulging in what's more formally known as 'power-levelling', meaning to power one's way through a game's content as fast as humanly possible. But you can also find gamer labels ranging from 'griefers' (who just want to cause exactly what the name suggests) to 'carebears' (who just want everything to be safe, friendly and risk-free).

## Twinks, twinked and twinking

Among hardcore gamers, this can lead to interestingly titled discussions on player forums. 'How many carebears are needed to feed a griefer?', for example, attracted considerable interest in the content of the space-themed massively multiplayer online game 'EVE Online' – although no definitive answer was found.[28]

# 27.
# Talking less about trees

Writing in 1978 about the future of language, the novelist Anthony Burgess speculated that one trend would be that 'language will be cut off from its roots in basic physical experience.' There will, he wrote, 'be a large everyday technical vocabulary to replace the old natural one'. But what we will end up with will be a 'language of the brain rather than of the body'.[29]

Burgess was looking to a time far distant from our own – the year 3000, in fact – but his point already resonates. Digital technologies are above all technologies of the mind, and the realm they most expand is one of mediated rather than physical experience: brain, not body, in Burgess's formulation.

It's interesting to look for evidence of the trend Burgess suggests – and a little tricky, not least because linguistic decline is harder to measure than the coining of new terms. One helpful resource, however, is Google's Ngram Viewer for its digital books service. This free online resource allows you to search for how frequently a word appeared within hundreds of thousands of books over the last few hundred years.[30]

The results are plotted as a graph and trace the percentage of total books within which the word in question was found, giving a sense over time of a word's increasing or decreasing popularity. Obviously enough, terms like 'computer' change from an insignificant level of usage before the 1940s to a steep, steady increase beginning around 1950 (and peaking just before 1990, with the arrival of other increasingly common digital terms: 'internet' and 'web' only really begin to show up on the graph at this point).

As an anecdotal test, I plotted on the graph the names of five of the most common British trees and birds, excluding those which – like 'swallow' and 'ash' – share their name with other unrelated words or names. This resulted in graphs tracing between 1900 and 2000 the occurrence of these words within English-language books: oak, elm, chestnut, hawthorn, yew; sparrow, blackbird, starling, finch, plover.

Every single one showed some degree of decline, perhaps suggesting – as Burgess argued – the lessening importance of the natural environment relative to the manufactured world; or at least a lessening interest among writers and readers in naming its specifics.

Also noteworthy is the rise in recent decades of natural and biological metaphors to describe the world's ever-more-complex digital systems: from app 'ecosystems' to the 'media landscape' itself: signs, perhaps, of an increasing tendency to treat technology itself as a kind of second nature.

Mine is hardly a rigorous experiment. It doesn't, however, need much investigation to discover an area of vocabulary in which perhaps the greatest growth in naming has taken place over the last century: that of machines, a field which has grown more thickly populated with every decade. At the start of the twentieth century, 'train' ruled the roost, overtaken by 'car' come the mid-1920s. 'Radio' briefly took the lead around 1940; but the supremacy of 'computer' since the late 1970s has left trains, cars, radios and television all trailing in its wake.

Today, in fact, 'computer' is a more common word in the corpus of English writing, as measured by Google, even than 'machine'. Not to mention – a fact that Burgess might have found of a piece with his predictions – there has been considerably more 'computer' since around 1980 than either 'tree' or 'bird' (although 'book', at least, remains more common than them all combined).

# 28.

# ZOMGs, LOLZ

Every day, billions of words of live conversation are exchanged online in a written format. Much of this pours out at close to the speed of speech, bringing with it one of the central new needs of written language in a digital age: words, phrases, symbols and expressions able rapidly to express those emotional nuances vital to informal conversation.

What are these nuances? Ordinary punctuation and vocabulary are pretty good at conveying conventional feelings of surprise, delight, enquiry, approval and disapproval. Push beyond these, though, and you soon hit a region of layered social complexities – of wit, irony, self-deprecation, knowing mockery, carefully calibrated sarcasm, and all of those registers that make informal interactions a minefield of feelings and intentions.

It's a minefield embodied in the existence of terms like ZOMG, a kind-of acronym whose first letter doesn't actually stand for a word, but which invokes the fact that people attempting to type OMG (Oh My God) on a conventional keyboard often hit the 'z' key while pressing the left-hand shift key for capital letters.

First encountered in online gaming culture in the mid-2000s, the typo soon became a new word in its own right, with the deliberate addition of the 'Z' supplying just the right amount of knowing sarcasm to denote faked surprise at something that ought to be obvious ('ZOMG, you're right, my computer doesn't switch on when it's unplugged!'). Today, the term can be used either to mock someone else's naive and stupid questions or to mock yourself, depending on context.

Such are the building blocks of a written conversational medium, and a field to which websites like the estimable UrbanDictionary.com are dedicated: terms and abbreviations with double- or triple-layered emotional resonances, suitable for banter of the most baroque kind.

Another nicely complex use of the 'internet Z' is the term 'lolz'. Originally just a plural form 'LOL' (Laugh Out Loud), lolz has gone on to become not only a fully fledged noun in its own right but the embodiment of a philosophy of online action. Doing something 'for the lolz' – or, if you're feeling still more amused, 'for the lulz' – conjures an often anarchistic delight in pranking and disruption, perhaps most infamously embodied by the hackers' collective LulzSec and its all-too-apposite motto: 'Laughing at your security since 2011!'

The process of finding new words for complex emotional states and traits is both wildly innovative and largely haphazard online. Yet it was prefigured, in 1983, by an entirely deliberate exercise. In their 1983 book *The Meaning of Liff*, subtitled a 'dictionary of things that there aren't any words for yet', Douglas Adams and John Lloyd used the names of places in England to describe particular feelings and objects. 'Abligo', for example, was 'One who prides himself on not even knowing what day of the week it is.'

Fittingly enough, several of Adams and Lloyd's words can today be found within the Urban Dictionary, where Adams himself is honoured with an entry that begins: 'Possibly the greatest English writer who ever lived . . .'

# 29.

# Happy pew pew to you

If you've ever wondered how to describe the sound lasers make when fired in the vacuum of space – or perhaps suspected that, being fired in a vacuum, they make no sound at all – look no further. The answer is 'pew pew'.

Onomatopoeias are among the oldest and most idiosyncratic of word types, although the term 'onomatopoeia' is disarmingly technical for something so elemental. A direct import from Greek, it combines *onoma*, 'name', with *poieo*, 'I make', and means simply a word that names the sound something makes. 'Woof' is a British onomatopoeia that names the sound a dog makes while, slightly bewilderingly, Arabic dogs go *hau*, German dogs go *wuff*, Hindu dogs go *bow*, Turkish dogs go *hev*, and so on.

Space lasers, however, go 'pew pew' in almost every language in the world, thanks to the unifying force of a particular twenty-first-century category: internet onomatopoeias. That is, sound-terms whose online distribution has ensured a near-global universality unheard of in all traditional examples of the type.

It's not all about lasers, though. Take a sound inextricably associated today with the eating noise made by cute cats (and other animals) in captioned memes: 'om nom nom'. Like many onomatopoeias, it's a phrase typically written in a plethora of different forms to generate a particular tone: from the 'omnomnomnomnomnomnom' of constant consumption to abbreviated versions like 'nom' and 'nom nom' for use in texting, casual chat, or as a kind of visual punctuation mark.

Google 'om nom nom' and you'll come up with close to nineteen million results – several times the paltry three million earned by 'pew pew'. Yet its origins predate memes and the internet entirely, being one of the favourite phrases of the *Sesame Street* television show character the Cookie Monster, featured in the very first show in 1967.

It may have begun on children's television, but it's the mad phonetic invention of the internet that has made 'om nom' into a global phenomenon – and fuel for a plethora of spin-off forms, largely originating in America. Eating especially delicious food might cause someone to have an 'omnomgasm'; while the food causing such an effect might be described as 'omnomalicious'. More obliquely, someone's very attractive mother might be described as an 'om nom mom'.

Verbal inventiveness is also only the beginning online when it comes to om nom, as sites like the eponymous www.omnomnomnom.com demonstrate. Pithily describing itself as a blog full of 'things edited in MSPaint to look like they're eating other things', it consists of selected photographs adorned with cartoon eyes, teeth and the universal caption 'om nom nom nom'.

Like many modern onomatopoeias, it's a phrase that's most at home in the company of an illustration – and the results can be weirdly beguiling. Especially as each image comes complete with a discreet text beneath, reading: 'If you're not saying "Om Nom Nom Nom" out loud at the same time as looking at these pictures then you're doing it wrong.'

# 30.

# Lifehacking

The term 'portmanteau' was first applied to words by the author Lewis Carroll – the pen name of the English mathematician and deacon Charles Dodgson – in his 1871 book *Through the Looking Glass*. At the time, a portmanteau simply meant a kind of suitcase (from the French *porter*, 'to carry'; and *manteau*, 'cloak'), but Carroll used it to describe 'two meanings packed up into one word'. That is, the practice of turning two different words into a single new word in order to describe something: 'fairy' and 'tale' becoming 'fairytale', for example.[31]

Today, portmanteaus remain a popular way of making new words, especially in and around the internet. One such word, complete with its own much-visited web address, is 'lifehacking' – as seen at Lifehacker.com – a term that neatly embodies not just the modern fondness for portmanteaus, but for steadily extending the compass of computing terms into non-technological fields.

The idea of lifehacking has been with us since 2004, arriving courtesy of British journalist Danny O'Brien and a session at the O'Reilly Emerging Technology Conference entitled 'Life Hacks: Tech Secrets of Overprolific Alpha Geeks'. Note that, at this stage, the word 'lifehack' itself didn't exist. O'Brien used two separate words in a novel combination in order to evoke a particular idea — that some of the most successful members of the new century's digital elite were using their skills not simply to hack computer systems, but to hack life itself.

What these alpha geeks were doing was devising productivity techniques that extended beyond the screen, into their management of all manner of daily tasks: from 'batching' similar tasks into set units of time, to using particular productivity software to schedule or co-ordinate work with others, or to establish particular rhythms of focus on attention-intensive tasks.

One popular technique among programmers for achieving focus is known as the *pomodoro* approach, and was invented in the 1980s by the productivity expert Francesco Cirillo, much of whose work is devoted to improving the working practices of software development teams. *Pomodoro* is the Italian word for tomato, and the technique takes its name from an iconic Italian tomato-shaped kitchen timer, which is used to break up work into twenty-five-minute periods of intense focus on a single task punctuated by five-minute breaks.

Each cycle of 25 + 5 minutes is defined as a 'pomodoro', and the idea of 'doing pomodoros' — usually in sets of four — has become standard practice among many programmers. It has also become a technique used outside of computing circles: a practice that's emblematic of the influence of the hacking mentality on life as a whole. This, more than anything else, is the essence of the term lifehacking today – a sign that the logic and techniques of managing our machines are increasingly becoming a model for successfully managing modern living itself.

# 31.

# The multitasking illusion

'Multitasking' is one of the supreme aspirational terms of modern mechanical magic. When you are multitasking, the assumption goes, you are embodying the epitome of contemporary productivity. To multitask effectively is, for some, a supreme professional and personal aspiration, because it means to be making the best possible use of that most precious quantity – time itself.

Etymologically, there's little that needs explaining: the word is a portmanteau of *multi* (Italian, 'many') and 'task' (ultimately derived from the Old French *tasque*, meaning 'tax' or 'duty'). What is interesting, however, is the fact that the word multitasking originated entirely in the context of computing – describing the scheduling of different tasks for processing – and was then subsequently extended as both a description and an aspiration for human actions.

Machine multitasking was, from the early days of computing, a necessity for allowing machines with often slow processors to perform more than one task simultaneously. With the right programming, it became possible for machines to perform more and more tasks without slowing down. The idea of 'threading' is also important in this process, a 'thread of execution' being defined in computer science as the most basic process that a computer's operating system can schedule. A computer, in other words, operates on a fundamental level as a kind of loom for the multiple threads of its processes, weaving them around each other in order to execute complex tasks.

This, clearly, is very different from the way in which the human mind operates – which is why the tendency to apply a machine-born word like multitasking to human activity is so interesting. Its implication is that to be engaged by multiple complex tasks simultaneously is desirable and efficient. Yet a mounting body of research suggests that, when it comes to any tasks requiring serious thought and attention, human multitasking is something of an illusion – and that what takes place when multiple demands are made of the mind is something closer to simple distraction.

One everyday example is the use of mobile phones when driving – a habit that's increasingly being banned in many countries on the basis of evidence of its dangers. In offices, however, it's still quite normal for people engaged in tasks every bit as demanding as driving to be constantly interrupted by email, calls, messages and the general accoutrements of the modern multitasked office and its technological systems.

Thinking differently is something that can be helped by using different words, and there's one particular phrase in this field that can be useful: Continuous Partial Attention. Coined in 1998 by the writer Linda Stone, it describes the state of constantly and largely unconsciously skimming across the surface of a large amount of different sources of information. Unlike multitasking, this isn't a business of focus and self-discipline so much as an attempt to go with the flow of information suffusion – and it suggests an important counterpoint to the belief that it's possible for people to, machine-like, juggle threads of activity seamlessly and without loss.[32]

# 32.

# The Streisand effect

Mass interactive media make for many unintended consequences, and the Streisand effect summarizes one of the most central: other people won't always do what you want them to, and efforts to dictate their behaviour can rebound rather messily.

The effect takes its name from beloved American singer and actress Barbara Streisand, who in 2003 sued a company and photographer involved in the California Coastal Records Project, on the grounds that one of their photographs was an aerial image in which her house could clearly be seen. The photograph, she argued, violated her privacy – an argument that was ultimately unsuccessful.

Far more significant than her lack of legal success, however, was the quite staggering lack of success Streisand enjoyed in attempting to preserve her privacy by suppressing the image of her home. As no less an authority than Wikipedia points out, before her actions drew attention to the image, it had been downloaded just six times from the website of the photographer who took it. Thanks to the legal case, however, over 400,000 people visited the site over the following month (and the image remains prominently reproduced across the web).

The term 'Streisand effect' itself was coined on the Techdirt blog in 2005, looking back on its coverage of the 2003 incident and at the way in which Streisand's actions had fuelled an internet-led movement doing precisely the opposite of what she wanted.[33] It's something that, as Techdirt's tongue-in-cheek 'Streisand Effect Hall of Shame' suggests, has become an increasingly familiar pattern thanks to the rise of global internet connectivity.

Other notable examples include the owner of the Washington Redskins American football team suing a newspaper over a satirical article criticizing him, thereby vastly increasing the attention given to the article, and a Scottish council banning a nine-year-old girl from blogging about her school meals, thus driving the story towards the front of every national newspaper in the country and culminating in an embarrassing about-face.

More seriously, the Streisand effect has also proved fully in evidence during the introduction of an unwelcome new word into popular consciousness in Britain: 'super-injunction'. Although the *OED* cites evidence for the term dating back to 1997,[34] it was coined in its current sense in 2009 by the editor of the *Guardian* newspaper, Alan Rusbridger, describing a legal injunction against reporting so ferocious that it even forbids reporting of the fact that an injunction against reporting has been issued.[35]

# The Streisand effect

Rusbridger first used the word in the context of the oil company Trafigura, which had used such an injunction to prevent any mention of the fact that it was prohibiting the reporting of an internal report on toxic waste dumping: a fact that was referred to in the British parliament, under the legal protection of parliamentary privilege, and subsequently circulated on the internet.

It was in 2011, however, that super-injunctions truly captured the popular imagination, thanks to attempts by British celebrities to prevent the reporting of potentially embarrassing elements of their public lives. In true Streisand style, these efforts earned many millions of words' worth of exposure on social media sites, highlighting along the way the impotence of some traditional legal approaches to personal privacy in a digital age.

For those wanting to control what the world thinks and knows about them, the Streisand effect embodies an important warning – that unintended effects are the rule, not the exception, of mass interactive media.

# 33.
# Acute cyberchondria

In Jerome K. Jerome's 1889 classic of English comic writing, *Three Men in a Boat*, the narrator describes his experience of reading a medical reference book in some detail. 'I plodded conscientiously through the twenty-six letters,' he notes, 'and the only malady I could conclude I had not got was housemaid's knee' – for every single other disease from A to Z, he recognizes at least some of the symptoms in himself.[36]

In the century since Jerome's gentle satire, self-diagnosis has swelled from being a neurotic habit of the middle classes to something far larger – a health obsession that the increasing availability of free information has swelled to epidemic proportions. In its digital incarnation, the trend is sometimes called 'cyberchondria' – a digital variant on the term 'hypochondria', and an etymologically unsound one at that, hypochondria being derived from the ancient Greek *hypokhondria*, 'upper abdomen', the region of the body from which feelings of general melancholy without a specific cause were once thought to originate.

## Acute cyberchondria

As a term, cyberchondria dates back at least to 2001, but it's only more recently that it has begun to achieve specialist as well as public usage. An academic paper entitled 'Cyberchondria: Studies of the Escalation of Medical Concerns in Web Search' appeared in 2008 from Microsoft's research division, for example, highlighting the ways in which the deluge of both medical information and disinformation online can fuel the impulse to self-diagnose.[37]

Perhaps a more intriguing phenomenon than mere cyberchondria, however, is 'Münchausen by internet' – a peculiarly contemporary variant on the mental illness Münchausen by proxy. Both conditions take their name from the legendary figure of the Baron von Münchausen, whose tall tales have provided a European archetype for the tendency to tell wildly exaggerated, attention-seeking stories.

Münchausen by proxy describes a psychological disorder in which someone seeks attention by telling false or highly exaggerated stories about someone else, giving it the 'by proxy' label. A parent suffering from the condition might, for example, repeatedly claim their child is unwell, even going so far as to simulate the symptoms of illness in that child.

The term Münchausen by internet was coined in June 1998 in an article co-authored by psychiatrist Marc Feldman,[38] but the behaviours it describes date back to the earliest days of the world wide web, and involve people seeking attention by telling false stories about their own illness and suffering online. A sufferer might, for example, construct an elaborate story about a terminal illness online, and seek out sympathy and advice via websites and forums devoted to helping genuine sufferers of the condition.

One early famous case of Münchausen by internet was perpetrated between 1999 and 2001, when an American blogger constructed a false identity called Kaycee Nicole, allegedly a teenager suffering from leukaemia whose writings gained a huge following. Kaycee was in fact the creation of a forty-year-old healthy woman called Debbie Swenson, who wove together online research and others' stories in order to create her fictional life – a quintessentially contemporary blurring of research, fact, fiction and online interaction.[39]

# 34.

# Casting the media net

Like many words associated with communications technologies, 'broadcast' is an old word that has gained entirely new meanings in the last century. First used in English in the second half of the eighteenth century, it originally meant to spread seed widely when sowing a field (that is, to cast it broadly).

These origins soon led to further metaphorical senses of the word, including the wide dissemination of information and ideas. By the early 1920s, it had fallen to 'broadcast' and 'broadcasting' respectively to provide a noun and verb for the young activity of transmitting information via radio.

Television took up the term upon its appearance, as did online services in turn. With interactive media, however, another related term had shifted its possibilities considerably: 'broadcaster'. Shortly after its launch, YouTube adopted the slogan 'broadcast yourself' as its motto, embodying a central truth of online media − suddenly, everyone with an internet connection and a computer could take part in something that had previously been the exclusive province of large corporations.

For historical reasons, the sense of the word broadcaster has shifted only gradually in the last few decades, still referring largely to either 'official' media presenters or corporations. What has happened, though, is the sundering of the second half of the word from its traditional formulation, and the birth of a whole spectrum of new kinds of '-casting' to describe new possibilities.

Perhaps the best known of all these new senses is 'podcasting'. The term first appeared in 2004, in an article for the *Guardian* newspaper by writer Ben Hammersley. Inspired by the dominance of Apple's digital music player the iPod, it described the new possibility of broadcasting via the iPod – or indeed any other digital music player – through the creation of downloadable audio files, individually known as 'podcasts'.[40]

Both the term and the practice took off, soon becoming an integral part of the digital music and broadcasting world – as well as spawning further linguistic variants, such as 'vodcasts' for video podcasts. Still more radical, however, is a different kind of '-casting' designed fully to exploit the new capacities of digital media. This is what's sometimes called 'lifecasting' – a verb conjuring the novel possibility of someone broadcasting information about almost every aspect of their daily lives online.

One strong candidate for coiner is the American pioneer Justin Kan, who in 2007 launched the website Justin.tv as a live record of his entire daily life relayed from a webcam attached to his cap. Under the title 'lifestreaming' – and in the noun form 'lifestream' – a similar concept had been discussed at least since 1994 by the American computer scientist and writer David Gelernter, who together with technologist Eric Freeman envisaged broadcasting someone's entire life live through a continuous 'stream' of digital images and documents.[41]

Although the technology involved is new, it's worth noting that the ideas underpinning lifecasting and its variants are ancient. The eighteenth-century philosopher Jeremy Bentham's notion of a panopticon (from the Greek *pan*, 'all', and *optikos*, 'relating to sight') embodied similar aims – a new kind of prison in which every single moment of inmates' lives could be observed, without their being able to see the observers.

Bentham imagined a round building constructed around a central 'observation house' and, though no prison was ever built that precisely matched his designs, it could be argued that mass interactive media have finally delivered a world in which his vision is being realized.

# 35.

# Bionic beings and better

From the start of the twentieth century until the 1960s, the word 'bionic' – an adjective formed from the Greek word *bios*, 'life' – existed only as a technical term within the study of fossils. To describe the 'bionic value' of different fossils was to quantify the enduring power over huge stretches of time of the characteristics of particular organisms.

Come the late 1950s, however, a very different sense for the word arrived, describing not the characteristics of living organisms, but the design of machines based on the study of biological systems, and the potential integration of machine and organic components. The coiner in question was medical doctor and United States Air Force colonel Jack E. Steele, known today as the 'father of bionics' thanks to his work from 1953 at the 6570th Aerospace Medical Research Lab at Wright-Patterson Air Force Base in Ohio, and who reportedly first used 'bionics' in 1958 to describe the copying of biological organs in the design of artificial prostheses and robots.[42]

# Bionic beings and better

Steele seems to have been unaware of the word's previous existence, and probably conceived it as a contraction of 'biological' and 'electronics'. The word itself remained an obscure item of technical vocabulary until the fateful intervention of mass media, in the form of the 1974 television series *The Six Million Dollar Man*. The series starred actor Lee Majors as military hero Steve Austin, injured while testing an experimental aircraft and subsequently reconstructed as 'the world's first bionic man': a part-man, part-machine super soldier. The series was a hit, and bred a spin-off series – 1976's *The Bionic Woman* – which helped cement the term in popular consciousness.

Interestingly, *The Six Million Dollar Man* was based on a novel that had already brought another young word into public consciousness: 'cyborg', meaning an organism that is part human and part machine. This term was also a child of the 1960s, having been coined as a contraction of the words 'cybernetic' and 'organism' in a 1960 article in the journal *Astronautics*, discussing the benefits of 'altering man's bodily functions to meet the requirements of extra-terrestrial environments'.[43] It was, however, Martin Caidin's 1972 novel *Cyborg* that boosted the term to the level of popular culture – a novel directly inspired by Jack Steele's work at Wright-Patterson Air Force Base.

Today, the increasing integration of electronic technologies into every aspect of human life have made bionics and cyborgs into iconic aspects of both science fiction and science fact — complete with ever-expanding linguistic subsets for each.

For example, a brain–computer interface (BCI), often called a mind–machine interface (MMI), is a direct communication pathway between the brain and an external device — and something that in recent years has achieved remarkable early results, including some of the first instances of rudimentary sight being restored to blind individuals via electronic retinal implants: tiny microchips that substitute for the functioning of damaged retinal cells.

As technology progresses, this newly literal integration of human experience with cutting-edge electronics is set to continue — bringing with it an increasing demand for new terms to describe the artificial, the enhanced, and the hybrid zone that lies between them.

# 36.
# Technological black holes

If someone begins to talk to you about 'the singularity', it's best to be prepared for a stretching philosophical experience. For 'singularitarianism' — as it's sometimes labelled — is a branch of futurist philosophy devoted to a simple enough idea. At some point, the hypothesis goes, artificial intelligence will reach the point where it exceeds human intelligence; a fact that will mark an evolutionary point of no return in the history of the human race.

It's this idea of a point of no return that gives the concept its name. The word 'singularity' — meaning a singleness of purpose — has been around in English since the fourteenth century, arriving from the Latin word of the same meaning, *singularitas*. For our purposes, however, the crucial notion is the twentieth-century concept of a gravitational singularity, more commonly known as a black hole: a point in the universe where so much mass is concentrated into so infinitesimally small a space that light itself is unable to escape.

Similarly, some thinkers reason, once the development

of artificial intelligence reaches a certain point of sophistication, intelligent machines will start to modify and improve themselves at an ever-increasing rate, bringing forth steadily more sophisticated machines within shorter and shorter periods of time. Human history will have, in effect, passed over the event horizon of a technological singularity, heralding the steeply accelerating arrival of a whole new kind of existence.

Perhaps the first person explicitly to describe such a technological singularity was science fiction author Vernor Vinge in the 1980s, whose early fiction explored the concept of super-intelligent future computers, and whose 1993 essay 'The Coming Technological Singularity' first formally outlined the concept of a near-future in which 'greater-than-human intelligence drives progress'.[44]

If Vinge was its pioneer, the most famous exponent of the singularity today is probably the American author and inventor Ray Kurzweil, whose 2005 book *The Singularity Is Near* offers a comprehensive guide to the putative near-future of superintelligent machines. Kurzweil's timeline puts the date for the singularity at 2045 — some fifteen years later than the date Vinge offered for its arrival back in 1993 — but paints a similar picture of what he calls 'a transforming event looming in the first half of the twenty-first century' brought about by 'the law of accelerating returns', with technological evolution functioning as an extension of biological evolution.[45]

Singularitarianism, as well as being something of a

mouthful, isn't the only iconic new word to be found within technologically inflected futurism. Perhaps the most intriguingly comprehensive label to be found here is 'trans-humanism'. The word first came to public attention in the 1989 book *Are You a Transhuman?* by the American thinker FM-2030 – whose 'name', as you might expect, was one he chose for himself as a reflection of his beliefs about escaping the conventions of human history.

Since then, transhumanism has grown into a global cultural movement, devoted to debating the future possibilities of technologically enhancing human minds and bodies. As you might expect, it's also a field whose neologisms suggest a certain state of mind, one of the finest of which is 'extropy' – an inversion of 'entropy', implying a steady increase in order and organization over time. As Kurzweil puts it, 'the list of ways computers can now exceed human capabilities is rapidly growing' – although, thus far, machines stubbornly refuse to help us find new words for what they're up to.

# 37.

# Google and very big numbers

The origins of the name 'Google' are a well-worn story, but worth repeating. The word 'googol' describes an enormous number – a one followed by a hundred zeros – and was thus used as the basis of the name for a company whose aspiration was to be able to index the seemingly infinite amount of online information (note, however, that a googol is a staggeringly huge number: significantly larger than the number of atoms within the observable universe, let alone the mere contents of the web).

As to why Google isn't called 'googol', explanations vary from the official version – that 'Google' was simply a deliberately playful variant on 'googol' – to the perhaps apocryphal story that this was the typo entered into a web browser when the company's founders were first looking to see what web addresses were free.

What's certain is that 'to Google' has today become a far more widely used word than 'googol' ever was, to the extent that it's virtually synonymous with the act of searching for information online. We should probably be grateful that the name of the original search engine built by Google's founders at Stanford in 1996 – 'BackRub' – was swiftly abandoned.[46]

Less well known is the story behind 'googol' itself. The mathematical word was born in 1938 in the most informal of contexts, thanks to the mathematician Edward Kasner's nine-year-old nephew, Milton Sirotta.

Kasner asked young Milton to come up with a name for a very large number that 'was not infinite, and therefore . . . had to have a name'. In response to his uncle's query, Milton coined both the word 'googol' and its partner, the still larger 'googolplex', which is ten raised to the power of a googol. A googolplex is so large that even writing it down in physical form – 1,000,000,000,000,000 . . . and so on – would require more space than actually exists within the known universe.

Not that the internet hasn't tried, of course. Wolfgang H. Nitsche's webpage 'Googolplex and other large numbers written out' allows you, for example, to run a program that might just generate a googolplex-worth of zeros on your screen, so long as you're prepared to run your computer continuously for around half a millennium.

Edward Kasner popularized both googols and googolplexes in his 1940 book *Mathematics and the Imagination*, co-authored with James R. Newman – although the latter is best known today thanks to the punning naming of Google's headquarters in California, the Googleplex.

The latter term, incidentally, was also used − in the form 'googleplex' − by the British author Douglas Adams in the fourth episode of the 1978 radio series *The Hitchhiker's Guide to the Galaxy*, in which the Googleplex Star Thinker in the Seventh Galaxy of Light and Ingenuity is named as an especially powerful super-computer (although not as powerful as the computer Deep Thought, which dismisses its rival as achieving mere 'pocket calculator stuff').[47]

# 38.
# Status anxiety

In his book of the same name, philosopher Alain de Botton observes that 'status anxiety possesses an exceptional capacity to inspire sorrow,' defining status in the broadest sense as 'one's value and importance in the eyes of the world'.[48]

'Status' is not a young word. It has existed in English at least since the 1670s, coming almost unchanged in form and meaning from the Latin *status*, meaning someone's standing, position or manner — a term that ultimately originated from a still older Proto-Indo-European word for much the same thing.

Literal and metaphorical status have always been closely entwined: an imposing physical stature is associated with power, while a ruler tends both literally and symbolically to be raised above others. Yet in just the last couple of decades, the internet has helped flesh out a newly important sense of the word: the broadcasting of precise information about what someone is doing, and what they wish to tell the world about themselves.

The origin of this sense is the use of 'status' in military and civilian settings to describe technical updates or alerts: standard jargon since before the birth of computing. What's changed in recent years is not the notion of status updates themselves in this sense, but rather the shift of something specialized and functional into a far more complex, compromised aspect of social interactions and standing.

It's a phenomenon that began in earnest with 'chat' programs like Microsoft's Messenger service, where an early innovation was the option to define your 'status' in order to indicate your availability (or not) for conversation.

As online chat programs grew more sophisticated, customizing your status became an increasingly significant form of online self-expression for users. It's an impulse that today has reached its apotheosis in Facebook and Twitter, and in two quite distinct domains: the provision of regular 'status updates' about whatever you're doing or feeling in a particular moment; and the display of comparatively fixed attributes like your 'relationship status'.

In each case, what's taking place is essentially a form of performance – and one that has earned a central place in digital culture. Posting status updates via Facebook and Twitter, a process sometimes referred to as 'micro-blogging', offers a kind of running commentary on your life and interests; while shifts within the broader categories of online status offer a point-and-click approach to social signalling, through which extensive networks of contacts can gauge their relative standing (and romantic availability). Facebook, for example, permits its users to self-identify not only as 'single', 'in a relationship', 'engaged' or 'married'; it also lists tick-box options for 'in an open relationship', 'divorced', 'separated', 'widowed' and – most iconic of all – the catch-all phrase 'it's complicated'.

Rarely has verbal self-expression been at once so effortless and so strictly boundaried: millions of words expended every moment in presenting one's own preferred face to other people; but constrained by the tightly defined interconnections of digital friendship, each one a link in an unprecedentedly literal approach to letting people know where they stand.

# 39.
# The zombie computing apocalypse

Along with vampires and werewolves, the stock of zombies has risen hugely in popular culture over the last few decades — a trend that has culminated, thus far, in creations as varied as a bestselling fitness app for smartphones entitled 'Zombies, Run!' (which encourages its users to go for runs while listening to an interactive narrative in which fictional zombies chase them) and leisure experiences like 'Zombie Shopping Mall', in which punters pay to take part in a live-action day of 'real' zombie fighting at a specially hired shopping mall, complete with actors and copious quantities of fake blood.

Etymologically, 'zombie' entered the English language via the creole spoken on the island of Haiti — a language born in turn from the mixing of colonial French with some of the African languages spoken by slaves brought to the island. In Haitian creole, the word *zonbi* denoted a dead body brought back to life by supernatural means — a word derived in turn from the Bantu language Kimbundu, where the term *nzumbi* described the spirit of the dead, as well as suggesting a connection to the animating spirit of a West African god in the form of a boa constrictor.

Thanks to Haiti's voodoo cult, zombies developed a rich mythos, including the notion of a 'zombie master' using magic and hallucinogenic drugs to keep both the dead and the living under a spell of mindless control. And, unlikely as it may sound, it's in this sense that the notion of zombies has most successfully crossed over from popular culture to the more specialized world of computing.

A 'zombie computer' is one that, unknown to its owner, has been infected with malicious software that allows a distant 'zombie master' to control it, along with potentially thousands of other machines similarly affected. The term arose in the early 2000s as a tongue-in-cheek way of describing the serious issue of what are more properly known as 'botnets' – networks of compromised computers acting as 'bots' (that is, automated robots) for a person or persons usually engaged in criminal or destructive behaviour. These controllers, similarly, are more properly called 'bot masters' or 'bot herders'.

Botnets are, today, a serious business for those operating on the darker side of the internet. Typically, they are used to send out spam emails – a sufficient number of compromised machines can be used to spew out millions of untraceable messages each day – or to mount so called DDoS attacks. Standing for 'distributed denial of service', these attacks entail instructing a zombie network of computers to connect repeatedly to a particular website or service, causing a huge increase in demand that almost invariably brings the target site crashing down.

The world's largest botnets can consist of hundreds of thousands of compromised machines, use of which may be hired out to the highest bidder by their operators, or made a formal aspect of a criminal enterprise. 'Night of the living dead computers' may not have quite the same ring as the film original – but it represents a far more imminent threat to human health and sanity, as well as an excellent reason for keeping an eye on your own system's digital health.

# 40.

# To pwn and be pwned

One consequence of the vast increase in rapidly typed communications that computers and the internet have brought is the emergence of the typo as a significant force for verbal innovation. It's a force that has been felt above all in those arenas which demand rapid typing under pressure – online games, for example, where players hastily bash out messages to one another while taking part in frantic onscreen action. It's also a trend that dovetails neatly with many online communities' insatiable hunger for private languages, codes, identifiers and neologisms.

Consider the term 'pwn'. One of its most notable features is that it's a word expressly designed to be typed, and read, rather than ever spoken aloud (if you do want to say it to someone, you can pick between pronunciations including 'pown', 'pone' or even 'pee-own', but in each case the effect won't really work unless it's written down as well).

To 'pwn' someone means to beat them completely and utterly, usually with a degree of humiliation involved. The term began its life as a common typo for 'own', as in the slang expression 'I totally owned you' – meaning, usually in the context of an online game, that I vanquished you so easily I didn't even break sweat.

Gaming and hacking communities have always overlapped, and 'pwn' in hacker jargon was soon being used to signify successfully taking over a target computer, or compromising and controlling a system. The variant noun 'pwnage' – as in the phrase 'there's some serious pwnage going on out there' – is just one further form of the word commonly used in both gaming and hacking. By the mid-2000s, 'pwn' had begun to take on a life of its own outside these subcultures, invoking any kind of humiliating defeat ('you got pwned').

This public emergence of 'pwn' is of a piece with the mainstreaming of numerous hacking and gaming terms, but perhaps the ultimate endorsement of its status remains within these fields, in the form of the so-called 'Pwnie Awards'. Founded in 2007, they recognize excellence (and humiliating incompetence) in the field of information security.

As you might expect, the categories of award to be found at the Pwnies are themselves a miniature taxonomy of hacking slang – from 'most epic fail' to special awards for 'mass' and 'epic 0wnage'. Note that the zero at the beginning of '0wnage' isn't a typo (although it may have begun its life as one) – merely another way of hacking the appearance of words themselves in order to make them more fitting tools of the tribe.[49]

# 41.

# Learning to speak l33t

When it comes to joining the digital tribe, the highest aspiration has traditionally become 'l33t' – pronounced 'leet', and standing for 'elite', denoting the top of the hacking, cracking, gaming and coding tree.

l33t is an idea that has steadily expanded its subsets and connotations with the expansion of the internet, but that began with a simple enough proposition. During the days of the early internet, filters were often applied to forums, user groups and bulletin boards to try to prevent the discussion of anything potentially illicit or offensive. In response, digital hackers and miscreants developed the habit of using alphanumeric characters as replacements for letters within words, together with deliberate mis-spellings, in order to elude censorship.

The use of a private linguistic code also dovetailed with hacking culture's obsession with status. Alongside 'l33t', perhaps the other single most famous term in the code is a label for the very opposite of the elite master of coding skills: the 'n00b'. Pronounced 'noob', the word emerged as an online variant of the American military slang 'newbie', indicating someone naive or inexperienced and thus without any technical skills.

Between these poles of expertise, both an entire vocabulary and alternative orthography has arisen, gradually making its way towards the mainstream as the place of computing in society has shifted. Orthographically – that is, in terms of spelling – l33t largely functions by using 'homoglyphs'. These are characters or letters that closely resemble one another, so can be substituted while leaving words largely recognizable: a '3' for the letter 'E'; a '5' for the letter 'S'; a zero for the letter 'O'; and so on.

Many of the other words in this book involve various l33t conventions and coinages, such has been its influence on digital vocabulary and culture. Remaining with orthography, though, it's worth noting that perhaps the most elite way it's possible to write l33t is entirely in numbers, in the form '1337' – or, sometimes, '1337 sp34k' ('leet speak').

Historically, greater and greater levels of orthographic obscurity have been associated with higher levels of expertise. Today, l33t is largely the victim of its own success and influence, and has become almost non-existent in terms of 'straight' rather than ironic, satirical or knowing usage. For those wishing to avoid detection online, mere scrambled letters are now a laughably crude approach – especially given the existence of extensive lexicons,

Wikipedia entries and glossaries of terms. The n00bs have taken over – and newer forms of obscurity must be sought elsewhere.

For those still wanting to taste the original tongue in all its glory, l33t is fully supported by Google as a search language in its own right, complete with buttons offering to '5h0w 0p710n5', let you 'st4Y Up 2 d4T3 on teH resULtz', or display '534R(h T1PZ'. Can you work out what each one means?[50]

# 42.

# Emoticons

The steady replacement of writing by typing has brought an inevitable homogeneity to the appearance of written language. Where once possessing an 'elegant hand' – that is, elegant handwriting – was an admired accomplishment, today we have software, styles and fonts to handle appearances for us. There are also more formidably flexible methods of personalization available.

First and most famous among all such techniques is the 'emoticon'. A term combining 'emotion' and 'icon', the word itself seems first to have appeared in the early 1990s. The objects it describes, however, date back considerably further.

Most online accounts name the 'father of the emoticon' as one Scott Fahlman, a computer scientist at Carnegie Mellon University, and its birth as 19 September 1982. This was the date on which Fahlman posted the following message to a computer science bulletin board: 'I propose the following character sequence for joke markers: :-) Read it sideways. Actually, it is probably more economical to mark things that are NOT jokes, given current trends. For this, use: :-( [51]

Fahlman's light-hearted suggestion addressed an important need in digital discussions: the need to indicate the emotional tone of a message, in order to ensure that enthusiasm wasn't misconstrued as sarcasm, a joke as something serious, and so on. Fahlman's two initial coinages would come to be known, respectively, as the 'smiley' and 'sad face' emoticons, and were the first to be specifically proposed for digital communication based on the standard ASCII (American Standard Code for Information Interchange) set of letters and characters.

They were, however, far from being the first such innovations. Over a century earlier, in the 1850s, Morse Code operators developed what was probably the first true emotional shorthand for electronic communications when a series of informal codes for standard sign-offs were introduced: 33 for 'Fondest Regards', 55 for 'Best Success', 73 for 'Best Regards' and 88 for 'Love and kisses'.

When it comes to typing, one of the first dates we can be sure of for emotional innovation is 1881, when an American satirical magazine called *Puck* published what it termed 'Studies in Passions and Emotions' using 'our own typographical line': four simple faces composed entirely of punctuation marks expressing joy, melancholy, indifference and astonishment. *Puck* was, it boasted, achieving through mere type the same effects as 'all the cartoonists that ever walked' – striking an early, if unwitting, blow for the primacy of typing as an inclusive force.[52]

Come the widespread use of typewriters by the start of the twentieth century, one standard abbreviation was the use of -) to express a 'tongue-in-cheek' comment, with other more personal innovations often ornamenting exchanges – a phenomenon echoed in the early days of computing via what was known as 'ASCII art'. That is, complex pictures and logos created entirely by using standard American Standard Code for Information Interchange characters, allowing the text-only realms of early discussion forums and bulletin boards to feature more than pure words (*Puck* would have been proud).

Like so much of digital culture, it was the widespread, interconnected use of computers that truly brought emoticons into their own. Since the two-face vocabulary of the early 1980s, they have now expanded and mutated into something verging on a distinct written language, complete with lengthy online dictionaries and dialects. Asian emoticons, for example, tend to depict upright rather than sideways faces: a tired face is (=_=) while confusion is expressed (?_?) Greater and greater obscurities also feature, often as elaborate gags or experiments in how far the form can be pushed. A person smoking might, possibly, be depicted (- -)y- in Japan; while an American Santa Claus could be *<|:-) – by which stage it's almost certainly easier simply to type the words out.

# 43.

# Getting cyber-sexy

While 'cyber-' is a prefix applied almost endlessly to digital versions of older activities and ideas, only one word has ever earned the right to be assumed in this context, and that's sex. To ask, in other words, 'Do you want to cyber?' is synonymous with asking 'Do you want to have cybersex?' And if the answer is yes, you won't even need to move away from your keyboard.

At its most basic, cybersex is a form of role-playing in which two people exchange sexually explicit messages online (what they do, or don't, actually do while typing these messages is up to them). As you might expect, this basic setup is almost as old as the internet itself, although the term 'cybersex' itself appears to date from the mid-1990s.

With the increasing sophistication of online technology, the possibilities for computer-mediated sexual activities have also grown more complex. Sexual interactions based on typing are still common, thanks to its unparalleled convenience and privacy: never before in human history have complete strangers so easily been able to exchange their most explicit, or obscure, fantasies with such anonymous ease or safety. The wide availability of webcams and microphones, however, has also opened the doors to far more revealing interactions.

While to 'cyber', then, is to engage in explicit chat, to 'cam' is to include visuals and sounds via webcam. If this is too self-exposing for your tastes, there's also an option at the other end of the spectrum known as 'mudsex' – a confusing word for outsiders, as no physical mud of any kind is involved. Rather, 'MUD' is an internet acronym for 'multi-user dungeon', and was coined as a description of the very first shared virtual world in the late 1970s. While the term itself has largely fallen into obscurity today, it remains in use as a description of simulated sexual activity between two people's avatars within a virtual environment.

Then we come to a field known, delightfully, as 'teledildonics' – whose etymology literally translates as 'the remote use of dildos', and which involves a combination of computer-controlled sex toys in arrangements such as a 'two-person teledildonics rig' (use your imagination) or the relatively young innovation of 'bluedildonics', which uses a Bluetooth connection to control a sex toy.

Such creations have been the subject of fevered science fiction imaginings since well before the technology existed to support them – indeed, teledildonics itself first emerged as a concept in the mid-1970s courtesy of sociologist Ted

Nelson, in the form 'dildonics', although the present term didn't hit popular culture until its appearance in Howard Rheingold's 1991 book *Virtual Reality*. More broadly, the category these kinds of human–machine–human interactions fit into is known as 'haptics', from the Greek *haptikos*, meaning something relating to the sense of touch. It's a field that encompasses not only pleasure and recreation, but the increasingly refined business of introducing physical feedback into digital interactions – potentially allowing, for instance, a doctor performing remote surgery to experience authentic physical feedback and resistance.

When it comes to sex and the internet, there's some truth in the old adage that words themselves are the most erotic medium of all. The massive field of 'fanfic' online – that is, fan fiction, which among other things has given the world the hit erotic novel *Fifty Shades of Grey*, first written as a tribute to the *Twilight* novels – is a case in point. Perhaps commonest of all, though, is the self-explanatory notion of 'sexting': sending sexually charged text messages. An activity that, for all its technological primitiveness, is arguably the modern world's leading form of erotic self-expression.

# 44.
# Slacktivism and the pajamahadeen

One of the most marvellously contemptuous coinages of recent times, 'slacktivism' means exactly what it sounds like: a 'slack' (i.e. ineffective) form of 'activism', as epitomized by the activities of net-savvy western youngsters perpetually willing to click on Facebook petitions, but almost pathologically unable to get out of their seats and take part in authentic political activity. That, at least, is the thesis behind slacktivism – a polar opposite to activism, typically used today as a stick for beating those who believe that new media alone can save the world.

The term itself was first used around 2001, with the excellent website Word Spy citing an article in *Newsday* in February that year as the term's first outing in print. Unlike some other digital terms, slacktivism is a word born specifically to describe online phenomena and to contrast these to 'real' actions. *Newsday* cited the example of campaigning emails which include the plea 'Forward this to everyone you know', but slacktivism's true home today is social media.[53]

## Slacktivism and the pajamahadeen

By contrast, another wonderful recent word describes those whose online actions more usually take the form of blog posts and extended investigations, and who may well have a real impact on their field. These are the 'pajama-hadeen': a punning combination of 'pajamas' and 'Mujahadeen' that conjures a vivid image (at least for me) of a fanatical campaigner clad only in their night-wear.

The term's origin is a classic tale of one man's insult becoming another's badge of pride. In 2004 America was in the throes of a controversy over whether George W. Bush — running for re-election as president that year — had completed National Guard service as he claimed. That September, the television show *60 Minutes* presented four documents which were critical of Bush's service. *60 Minutes* claimed the documents were authentic, but soon came under extensive criticism from bloggers and discussion forums — criticism that was then picked up by the mainstream media.

Defending *60 Minutes*, a network executive witheringly contrasted the professional journalists involved in making the programme with 'a guy sitting in his living room in his pajamas writing' online. It was a comparison his opponents leapt upon — not least because many of the points the bloggers raised proved to be valid — and adapted into a new word extolling the new political power of people simply sitting at home behind their computer screens.[54]

# 45.
# Gamification and the art of persuasion

Buzzwords come and go, but one that – perhaps unfortunately – seems to be here to stay is 'gamification'. Etymologically, there's not much of interest in it beyond the literal meaning, 'to make something like a game'. Unpacking the term and its associated vocabulary, though, offers some intriguing insights into current culture.

The word has been in vogue only since the late 2000s,[55] and reflects a sea change in the world of digital gaming. Where once playing video games was largely seen as the pursuit of teenage boys, the presence of more than a billion people on social networks like Facebook, and the ownership of powerful smartphones by hundreds of millions, has created a massive audience for entirely new forms of 'social' and 'casual' gaming.

From the bestselling Angry Birds games – which have now been downloaded over a billion times, and which invite players to catapult cartoon birds into cartoon fortifications built by evil pigs (natch) – to new variations on old classics such as Scrabble, digital play has found a whole new audience and level of integration into many lives. With this, inexorably, has come the belief that lessons can be taken from some of the successes of the games industry and applied to other fields.

Perhaps the most common current use of gamification occurs when a business sets out to 'gamify' their online activities. That is, they attempt to apply lessons from the realm of video-games to consumer interactions with their product or service in order to achieve higher levels of engagement, enjoyment, loyalty, and so on. Typically, this involves the use of 'experience' points, special 'badges' and 'achievements' for performing certain tasks, and a strong 'social' element encouraging co-operation and competition among users.

All of which are tactics that would have been familiar in the 1950s to psychologist B. F. Skinner – the father of 'operant conditioning', an approach to learning that seeks to modify behaviour through calibrated rewards and incentives. What's new, though, is the degree to which such tactics can be automated, refined and expanded exponentially in scale thanks to online technologies. These principles have perhaps reached their apogee in the young field of 'advergaming', or games constructed explicitly to function as adverts for brands or products. From Burger King to Hollywood movies, digital play has become a potent marketing weapon.

As well as mere profit, there has been considerable growth in the last decade in the seemingly oxymoronic field of 'serious gaming', which seeks to use digital games of all kinds for purposes ranging from education and awareness-raising to training in particular technical or management skills. All of this overlaps in turn with the burgeoning development of 'persuasive technologies' – another slightly unsettling phrase that describes the engineering of technological systems explicitly to promote certain kinds of behaviour or belief.

It's a peculiarly contemporary form of the older science of behavioural modification, and one that to some ears has a dystopian twang. What's inescapably true, though, is that the technological systems saturating our lives increasingly exert powerful pressure on our actions and thoughts alike – and that even taking refuge from the world in the delights of a well-made video game is an act increasingly loaded with both economic and political baggage.

# 46.

# Sousveillance

Etymologically, surveillance means 'watching from above' – from the French words *sur* ('above') and *veiller* ('to watch'). It's a term that has been used in English since the start of the nineteenth century, and refers to the covert monitoring of a situation by an individual or institution.

The far more recent word 'sousveillance' inverts all this, taking its first four letters from the French word *sous* ('under'). It's a knowing reversal of the traditional term, and one that captures an important contemporary phenomenon: the monitoring and recording of a situation not by a covert external authority, but by someone actually participating in it, usually via a portable device like a smartphone or webcam.

This idea of 'watching from below' is a core element of the democratization of media technologies – that is, the process that over the last few decades has seen recording and broadcasting devices become both affordable and increasingly universal. The word 'sousveillance' itself emerged well after the impact of this change first began to be felt, and appears to date to a 2002 coinage by Steve Mann – one of the pioneers of wearable computers and cameras.[56]

Mann defined sousveillance as 'watchful vigilance from underneath', emphasizing its anti-establishment credentials, inspired by the experience of pointing his own recording devices at surveillance cameras recording him – an act he originally termed 'shooting back', and which was the title of a video work he filmed using precisely that process of filming those engaged in filming him.

In 1998 Mann also launched in Canada what he at that time called 'National Accountability Day' – marking out 24 December, the busiest shopping day of the year, as a time for 'shooting back' that has gradually become a global phenomenon. In 2002 it took on the label 'World Sousveillance Day'.

Perhaps Mann's most radical gift to both vocabulary and the field of sousveillance itself is the device called an 'EyeTap'. A tiny camera worn in front of the eye, it is able both to record exactly the same image as the person themselves is seeing, and to superimpose on their vision a computer-generated image. EyeTaps are in turn a powerful tool in the practice known as 'Cyborglogging' – that is, the construction of 'cyborg logs', or blogs created by people who are using wearable computing and recording devices in order to function as 'cyborgs' (a word itself explored earlier in this book).

The idea of always-on wearable cameras and computing

took another leap forward in 2012 when Google gave the first public demonstration of its 'Glass' technology – a pair of glasses containing a fully functioning networked computer, camera and microphone. With the increasing affordability and power of such technologies, we're entering a realm sometimes described as 'coveillance' – that is, the parallel process of wired-up citizens all watching each other.

This kind of mutual broadcasting and recording is unlikely to be welcomed by everyone, a fact that led in around 2005 to the coining of the notion of 'equiveil-lance', described by Mann and his colleague Ian Kerr as the possibility of finding a happy equilibrium between surveillance from above and sousveillance from below. It's a proposition towards which Kerr confesses feeling distinct 'ambi-veillance', not to mention an unfortunate fondness for awful puns.[57]

# 47.

# Phishing, phreaking and phriends

To those in the know, swapping an 'f' for a 'ph' at the start of a word is a sure indicator of digital provenance. The prototype, however, predates even the internet, drawing on an innovation from the very earliest days of electronic hacker culture: 'phone phreaking'.

It's generally thought that phone phreaking began in America in 1957, when Joe Engressia, a blind seven-year-old child with a remarkable gift for musical pitches, discovered that by whistling a particular tone into his telephone, he could cut off a phone message from an unconnected number. Young Joe had unwittingly discovered that a 2600 Hz tone had been built by the telecoms company AT&T into its system as an automatic switch, for internal company use.

Using this particular tone automatically made the telephone switching system act as though a call had been finished, leaving the line open and — as Engressia and those who came after him soon discovered — the person holding the phone in a position to make international and long-distance calls free of charge.

## Phishing, phreaking and phriends

Joe Engressia's talent – and the skills of his friend, John Draper – first came to public attention in a 1971 article for *Esquire* Magazine by Ron Rosenbaum that also seems to have coined the term 'phreaking' itself.[58]

Previously to this, Engressia and his comrades had described themselves as 'phone freaks', a phrase evoking both their outsider status and their obsessive interest in the intricacies of the phone system (the term 'freak' itself had been in use as a term for someone extremely keen on something since the first decade of the twentieth century, apparently originating in the phrase 'Kodak freak' to describe a committed early cameras expert).

Rosenbaum compressed the two words into one neat term – and in the process offered the world not only an iconic word, but a variant spelling that would come to be associated with wider hacker culture far beyond the fooling of telephone exchanges, thanks not least to the publication in 1985 of what would become one of the world's most influential hacker e-magazines, *Phrack*. This institution continues to this day at the site phrack.org, and took its name from a combination of the words 'phreak' and 'hack'.

Two of the most famous of Engressia's successors were the co-founders of Apple, Steve Wozniak and Steve Jobs, whose exploits in the 1970s – including a legendary prank call to the Vatican, purporting to come from Henry Kissinger – were probably inspired by the original *Esquire* piece. Orthographically, however, perhaps the greatest legacy of phreaking lies in the distinctly more unpleasant field of 'phishing': a form of email scamming in which the scammers are effectively fishing for gullible recipients by sending out purportedly legitimate emails asking for personal and banking details.

The word 'phishing' itself appeared around 1995, its first recorded mention being as part of the hacking tool AOHell, a program designed to help its users perform various kinds of hacking via AOL (America Online). One of these was a tool that would automatically send fake messages to other AOL users, purporting to be a 'security check' from AOL itself and requesting username and password details – a crude approach, but one that proved sufficiently successful to define both the language and the techniques of countless subsequent subterfuges.

# 48.
# Spamming for victory

Of all their bequests to popular culture, perhaps the most enduring gift of British comedy series *Monty Python's Flying Circus* may prove to be a digital one: the term 'spam', a catch-all description for the more than one hundred billion items of nuisance email sent every day.

The Monty Python episode in question was first broadcast in 1970, and featured a sketch known simply as 'SPAM': the brand name used since 1937 by the Hormel Foods Corporation as a contraction of the phrase 'spiced ham' to describe its canned, precooked pork product. Set in a British cafe with almost every single item on the menu featuring spam – often several times over – the sketch culminated in a chorus of mysterious Viking warriors drowning everyone else's voices out by ceaselessly chanting the word 'spam' (which also went on to infest the show's closing credits).

This satirical indictment of British culinary monotony began to take on its second life during the early 1980s, when – before the advent of the world wide web – the

early internet was dominated by message boards and
bulletin board systems. A delight in mischief was always
a prominent a feature of digital culture, and those who
wished to derail discussions developed a habit of copying
out the same words repeatedly in order to clog up a debate.

Inspired by Monty Python, the word 'spam' itself and
other lines from the sketch proved a popular way of doing
this – an effective tactic when some internet connections
were so slow that loading hundreds of letters could take
a considerable amount of time. In online discussions and
early text-based games, the term 'spamming' soon came
to describe any process of drowning out 'real' content with
repeated, low-quality words or other interference.

It was the global growth of electronic mail, however,
that provided opportunities for spamming on a hitherto
unimaginable scale. Instant, free and unlimited in scope,
email was the perfect vehicle for speculative mass communications: a trend usually thought to have begun in
1978, when one of the first recorded junk emails was
sent. It came from a Digital Entertainment Corporation
marketing representative, and was aimed at almost every
single person with an email address attached to the early
internet on the west coast of America (around 600 people
at the time).[59]

The term 'spamming' had yet to be used in 1978, but
the pattern was a more-than-familiar one. Receiving unsolicited communications via the postal service was already
a long-established practice by the time the phrase 'junk

mail' appeared in 1954. In America, unsolicited commercial messages were being sent over the telegraph system from the mid-1860s, while a number of disgruntled British members of parliament received such messages at their doors in 1864 as an advertisement for the opening of 'Messrs Gabriel, dentists, Harley-street, Cavendish-square'.

And so the inundation has grown alongside communications technologies to biblical proportions, extending today not only to emails, but to everything from 'spam blogs' (populated automatically with commercial links) to 'spam bots', which join chat rooms and online games, automatically generating links and enticements. So useful is the term as a descriptor that its meaning continues to widen, encompassing everything from unsolicited marketing calls ('spam marketing') to printed literature.

With pleasing symmetry, it's also now common practice to refer to desirable emails as 'ham' – and even to ask email senders to use a 'ham password' to confirm that they are, in fact, a real person rather than a computer. More inventive means for fighting the spam hordes (both technologically and linguistically) include digital 'honey-pots', which simulate email systems ripe for hijacking by spammers but are in fact intended simply to waste their time; and 'tarpits', which are email systems designed to slow the rate of spam arrival to a crawl. Electronic spam, though, is here to stay (as is the foodstuff, which still sells over a hundred million cans a year around the world).

# 49.
# Gurus and evangelists

How seriously should you take someone who has the job title 'social media guru'? It's not my question (the original was posted on the online questions and answers repository Quora, with answers tending towards 'about as far as you can kick them'), but its very existence makes an interesting point. Digital technologies have become one of the great secular religions of our age.

In Sanskrit, 'guru' means a teacher or master. The word entered English from Hindi at the start of the nineteenth century, still carrying its original sense of a teacher or priest – and seemingly the first person to whom it was applied in the sense of 'leading expert' was the Canadian pioneer of media studies Marshall McLuhan during the second half of the 1960s.

Culturally, the 1960s were a period during which Eastern mysticism was very much in vogue – thanks not least to the prominence of the Beatles' own personal guru, the Maharishi Mahesh Yogi. With his two books *The Gutenberg Galaxy* (1962) and *Understanding Media* (1964),

## Gurus and evangelists

McLuhan was positioned as perhaps the world's leading authority on the current direction of popular and media culture.

From the mid-1960s onwards, a number of American newspapers and publications began referring to McLuhan as a 'guru' of, variously, popular culture, mass communications, media, and even a cult: a usage that may well have been begun by none other than the author Tom Wolfe, writing in 1965 for the *New York Herald Tribune* in a series of pieces called 'The life out there' – one of which he devoted to 'this man, this pop Guru McLuhan' under the title 'What if he's right?'[60]

Such labels fitted both McLuhan's influence and his style, which was eclectic and could border on the mystical, bequeathing the world axioms including 'the medium is the message'. A charismatic and popular speaker, McLuhan was the prototype of the social media guru – although it's telling that the word itself was chosen for him by others, usually with a sardonic or sceptical undertone, rather than a label he chose for himself.

Today, so-called social media gurus are fixtures in the digital landscape; as, too, are another breed of religiously inflected technology enthusiast, the digital evangelists. Sometimes known as technology evangelists, the term adheres closely to its etymological roots, which lie in the Greek word *euangelizesthai* meaning 'bringers of good news'. Where the good news that the Christian evangelists brought was the Gospel, however, the good news that technological evangelists bring tends to be about why a particular digital product or service deserves your attention.

This notion of attention and loyalty is a vital one in the intensely competitive realm of digital services, and one that has made the role of evangelists a central one in many tech companies – to the extent that it serves as an official job title at some firms (Google has boasted a 'chief internet evangelist' among other senior roles at the company). As the word suggests, something more is at stake than simply sales and marketing, not least because of the level of belief – indeed, of faith – embodied in that most crucial of all contemporary decisions: which mobile phone should I buy?

# 50.
# CamelCase

What do YouTube, eBay and Leonardo DiCaprio have in common? All three represent a typographical phenomenon sometimes known as 'medial capitals' – or, more poetically, as CamelCase. All of them, that is, are written with some capital letters mixed into lower case letters, without spacing, in order to make them easier to read and pronounce.

It's a logical approach that will be familiar to chemists, who have been using mixed case letters for clarity in chemical formulae since the Swedish chemist Berzelius first suggested it in 1813. Iron oxide is, for example, written 'FeO', combining the symbols for iron 'Fe' and oxygen 'O' without a space between them, to indicate their combination.

Chemicals may have helped start it, but CamelCase really came into its own with the birth of computer programming via typed programming languages in the 1970s. For many elements of early programming languages – such as the names of files and variables – spaces were not permitted as characters, meaning words and terms had to be run together while typing.

Coders compelled to type line after line of highly complicated text found that using capital letters within these strings of letters was an extremely useful way of keeping track of commands. Typing 'PrintScreen' is, for example, easier to read at a glance than typing 'printscreen' — especially if you're looking over hundreds of lines of code written by someone else.

There are, technically, two different types of CamelCase: UpperCamelCase (UCC), in which the first letter of each word is capitalized, and lowerCamelCase (lCC), in which the first letter is lower case. All of which would perhaps be of little interest to anyone other than the most dedicated of linguistic researchers, were it not for the coming of the world wide web.

Spaces are not acceptable as part of a web address — and, as the web grew through the 1990s and 2000s, this simple fact suddenly put everyone in the position programmers had occupied. If you wanted to be able to distinguish between the different elements of often-complex website names and addresses, it made sense to use CamelCase; something that was doubly true if you were a company wanting your name and web address to be both identical and as legible as possible.

It's important to note that the web didn't invent CamelCase as a style of corporate branding. Perhaps the first companies to deploy it were the rival film format brands CinemaScope and VistaVision in the 1950s. It is, however, fair to point to the web as a massive popularizer of the technique – something that has also seen it become fashionable as well as functional, perhaps because of the association with leading technology brands from eBay to YouTube via iPods and PayPal.

The charming term CamelCase itself, though, does seem to be a child of the digital age. Its likely origin is a 1995 Usenet posting by one Newton Love, who offered this as his pet term for the typing convention thanks to 'the humpiness of the style'. As so often online, the whimsically appealing won out over the merely traditional or authoritative, and the term seems set to remain a permanent part of typography.[61]

# 51.

# The Blogosphere and Twitterverse

Most words ending in '-sphere' describe aspects of the natural world. The Latin word *sphaera* originally described a globe or ball, as did the Greek word *spharia* on which it was based, but these terms were also used to describe the 'spheres' through which the moon, sun, planets and stars appeared to move around the earth.

We no longer believe the universe revolves around the earth, but we still talk about the atmosphere and its subdivisions when describing the layered air surrounding the earth, from the troposphere touching the earth's surface to the exosphere six hundred kilometres above it. Similarly, the 'biosphere' describes the parts of the Earth and its atmosphere within which life exists.

It's natural enough, then, that the suffix '-sphere' has over the last two decades become a standard form for those information systems that also girdle the earth, albeit metaphorically speaking.

First among these is the 'blogosphere' – a word initially coined, like many fine digital terms, as a joke. The coiner was the writer and pioneering blogger Brad L. Graham, who on 10 September 1999 posted an entry on his blog, the BradLands, speculating about the booming future of the medium. 'Is blog- (or -blog) poised to become the prefix/suffix of the next century?' he asked. 'Will we soon suffer from (and tire of) blogorreah [sic]? Despite its whimsical provenance, it's an awkward, homely little word. Goodbye, cyberspace! Hello, blogiverse! Blogosphere? Blogmos?'[62]

Coined partly in tribute to the older notion of a 'logosphere' (from the Greek *logos*, 'word,' and meaning the global realm of discourse or intellectual discussion), 'blogosphere' rapidly migrated from being a tongue-in-cheek coinage to something neatly slotting into an empty linguistic space – a term fit to describe the interlocking global network of blogs, and the quasi-communal atmosphere of communications across them.

The term 'blog' itself was young when blogosphere was coined. A compression of the phrase 'web log', it dates from the mid-1990s, although it was the launch in 1999 of the online service Blogger that first brought blogging as an activity to general web users as well as early-adopting experts. Similarly, relatively little time has elapsed between the public success of other new social media and communications services and the coining of new terms for their community of users.

The micro-blogging service Twitter, for example, was launched in 2006, and as it grew in popularity the coinage of 'twitterverse' swiftly followed — a collective description of the world's tweeting that has proved more popular than the alternative 'twittersphere', although both terms are used.

Within each realm, a host of user labels and subdivisions exist, some of which have become internationally recognized words in their own right. Twitter followers are often known as 'tweeps', while gatherings of twitter users can be called 'tweetups'. Both 'blogaholics' and 'tweetaholics' are those who indulge rather too compulsively in their preferred online forms of self-expression — of which 'blogorrhea' and 'twitterrhea' respectively may be the symptoms, signifying far too much sharing of far too much information with a largely indifferent world.

# 52.
# Phat loot and in-game grinding

Like hacking, online gaming features prominently in most accounts of digital innovations in vocabulary, thanks in large part to the fact that millions of people playing and competing together in cheerful anarchy tend to innovate at a frantic rate.

Much of gaming's vocabulary is also highly functional: fit for fast typing during exchanges of virtual fire, and matched to the particular technicalities of a game genre. As in hacking culture, too, fluency in acronyms and jargon is a mark of hard-won experience – a way of differentiating the elite from mere initiates, and of signalling membership of particular tribes.

One of the broadest jargon categories in gaming is that of general-purpose acronyms, defining common game mechanics and characteristics: from DOT ('damage over time') to FPS (either 'first-person shooter' or 'frames per second', depending on context), NPC ('non-player character'), DPS ('damage per second') and dozens more.[63]

Then there are the words that describe standard activities within many games, and especially the Massively Multiplayer Online genre — or MMOs for short. Most of these activities relate to progression, combat, rewards and hierarchies. The verb 'grinding', for example, describes a player deliberately slogging their way through repetitive activities in order to progress (and has a nicely visceral sound, coming direct from the Old English *grindan*, 'to scrape or rub together').

When it comes to what they're grinding, the standard description of computer-controlled enemies encountered within games is 'mobs' (short for 'mobiles', meaning mobile objects, and originally a programming term used in 1980 by one of the creators of the very first shared virtual game-world, Richard Bartle[64]); and the standard term for those numerous computer-controlled mobs that pose no real challenge and offer no great rewards is 'trash'. String all this together, and you have a sentence worthy of admission to the middle echelons of the gaming hierarchy: 'Grinding trash mobs sucks.'

The in-game rewards of play, meanwhile, tend to be termed 'loot' — or, in exceptional circumstances, 'phat loot'. But simply grinding ordinary mobs doesn't earn this kind of reward. For higher-end players, it tends to be all about highly co-ordinated 'runs' into the most challenging parts of a game; or 'farming' a particular area in the hope of 'rare spawns' — that is, the infrequent generation by the game's algorithms of unusual enemies whose slaughter may bring dazzling rewards.

## Phat loot and in-game grinding

Then there comes the more technical vocabulary used by players and critics alike to dissect games, and to critique the medium with a depth and precision once reserved for films or works of literature. Foremost among a game's desired traits is 'balance' – a balanced game being one in which, whatever a player chooses to do as they progress and explore, the levels of challenge and reward remain in a carefully defined 'sweet spot' of enjoyment rather than becoming either boringly easy or frustratingly difficult.

Achieving such balance is closely related to defining the 'difficulty curve' of a game, as well as engineering its 'mechanics' correctly by ensuring that the virtual physics of its environment and the ways in which players are permitted to act are all seamless and satisfying. Shoddy game design might permit 'exploits', allowing players to progress or gain assets in unintended ways. To leave the 'play experience' feeling 'unbalanced', though it may sound mild, is one of the most serious indictments of design failure any true gamer is likely to utter.

# 53.
# Meta-

In ancient Greek, the preposition *meta* meant 'beside', 'beyond' or 'in common with'. It was this sense of being 'beyond' something else that led to its use in the phrase *meta ta physika* ('coming after the Physics') to describe those works of Aristotle's which were usually placed after his works on physics when all his writings were gathered together.

This led in turn to one of the more significant verbal confusions in philosophical history, as Latin writers began to use the label *metaphysica* as the actual title of Aristotle's major philosophical works, rather than simply a description of the location of these books.

'Metaphysics' soon became enshrined as the branch of learning concerned with matters 'beyond physics' – and so the near-mystical sense of the word was born, and with it the birth of 'meta-' as a prefix denoting not simply location, but anything concerned with the most fundamental or underlying nature of a field.

It's a term that has existed in more rarefied English

vocabulary for centuries, but it began truly to come into its own with the rising interest in so-called 'meta-theories' at the start of the twentieth century (that is, theories about the nature of other theories). This usage of the term was inspired by mathematician David Hilbert, who in 1904 announced his intention to demonstrate the consistency of logic and arithmetic by developing a theory of 'meta-mathematics'.[65]

The great popularizer of the term 'meta' for contemporary purposes, however, was the American writer, physicist and mathematician Douglas Hofstadter, whose 1979 book *Gödel, Escher, Bach* explored – among many other things – the idea of self-referential and self-generating theories, equations and frames of reference.

Hofstadter's work has continued to be hugely influential, not least because his ideas have proved a perfect match for the endlessly self-referential loops of the internet itself. It's probably thanks to Hofstadter that the prefix 'meta-' has today become a fully independent adjective in its own right, typically used online to describe something self-referential or knowingly oblique. A satirical six-word 'autobiography' of Hofstadter was even devised by cult web comic xkcd, paying tribute to his reflexive theories: 'I'm So Meta, Even This Autobiography . . .' (you have to read the first letter of each word to appreciate the depths of this particular meta-ness).[66]

In parallel to this broad cultural sense, 'meta' is also frequently found today in the important field of 'metadata'.

It's a vital discipline in an information-rich world, and most commonly describes the information 'tags' attached to data sets in everything from blogs and archives to vast experimental data sets.

It offers, in other words, a kind of floating, higher layer of information about information itself – and with it the possibility of a whole new phase of self-organizing digital culture. We might have to wait longer than you'd think for this future to arrive, though Hofstadter himself surmised in his eponymous law of expectations: 'Hofstadter's Law: It always takes longer than you expect, even when you take into account Hofstadter's Law.'

# 54.
# TL;DR

Words are cheap online, and sometimes there are simply too many of them. It's for such circumstances that the five characters 'TL;DR' have developed as a staple of online discussions. They stand for the phrase 'Too Long; Didn't Read', and are traditionally deployed as a response to an excessively long piece of comment or argument in an online debate (or as a humorous way of asking someone to stop waffling and get to the point).

'TL;DR' is an interesting acronym, not least because it's one of the very few to contain a semicolon – a hint at its likely origins among the ranks of editors on Wikipedia and members of other less high-minded online forums like FARK, where it first began frequently to be used around 2003.[67] One unusual variation on TL;DR is an animated image of a teal deer – sometimes used in online postings due to their similar pronunciation – but the ethos it embodies today is more often expressed both without the 'official' semicolon and in deliberate haste.

At the opposite end of the rhetorical scale to TL;DR is

a far older acronym devoted to the idea that someone should read the relevant text in depth before joining in a debate – a sentiment more often expressed as 'RTFM', or Read The Fucking Manual. It's thought to have arisen among technicians in the US Air Force during the 1950s, or perhaps even earlier during the Second World War, but didn't reach print until the end of the 1970s, when it was used (without being spelled out) as an insiders' joke on the contents page of a manual for the software library system LINKPACK.[68]

RTFM remains in widespread use online, together with a more recent coinage also intended to suggest that people should look something up before asking an obvious question: 'Google Is Your Friend,' or GIYF for short. This relatively polite phrase also has its less civilized variants, including the pithy JFGI ('Just Fucking Google It'), whose profanity is surely indebted to RTFM and which comes complete with its own  website at http://justfucking-googleit.com/ for those who really don't understand.

Both saying too much and understanding too little can be fatal flaws online – although it's worth noting that the culture of sites like Wikipedia has also bred a particular emphasis on establishing efficient protocols for explanation. Hence one of the most pleasing linguistic popularisations of a culture in which almost anything can be looked up by anyone: 'disambiguation', meaning 'the removal of ambiguity'.

Although dating back to the early nineteenth century,

the term has come into its own online thanks to the notion of 'disambiguation pages', within which things sharing the same name are differentiated so that those typing them into a search box can find the relevant result. Type 'top' into Wikipedia, for example, and the disambiguation process will ask you whether you mean clothing, a DC Comics supervillain, part of a ship's rigging, an album by The Cure, a short story by Kafka, a sexual role, a US Marine Corps or Army rank, a topological space, the largest semisimple quotient of a module, the greatest element in a partially ordered set, a quark, one of two places in Azerbaijan or a Romanian river, a Unix program, a data type in computer science theory, or a Malaysian/Indonesian Muslim extremist. Or just a spinning toy. Clarity is all.

# 55.
# Apps

Etymologically, 'application' means 'bringing something to bear on something else', from the Latin *applicare*, meaning to join or attach. The word has been with English since the early fifteenth century – and has come to mean both the act of applying oneself dedicatedly to a task, and subsequently the business of applying for a job or position.

Its most common usage today, however, lies in the digital realm; a sense of the word that in many ways harks back to its oldest meaning. For in computing, the term 'application' was originally shorthand for the phrase 'application software' – this being software specifically designed to help a user with a particular task, such as word processing or creating a database.

Using application software means bringing a specially designed program to bear on a particular task in order to make it easier: something that in the early days of computing was not always common practice, and where computer users might be expected to solve most of their problems from scratch using only a computer's basic operating system.

It was the dawning age of personal computing in the 1980s that saw application software first come into its own, facilitating the use of new machines for a wider public than the first generation of enthusiasts who had bought microcomputers in the 1970s. Perhaps the greatest shift in the usage of the word itself, however, has come alongside the similarly momentous shift in hardware and software associated with the advent of smartphones and tablets since the second half of the 2000s.

In 2008 Apple launched its eponymous 'App Store', designed to create a digital marketplace for the sale and distribution of applications for devices such as its new iPhone. The term 'app' was carefully chosen. Since the early 1990s, it had been an accepted shortened form of the term 'application'. This was, however, one of the first times that the abbreviation was itself enshrined in an official name, bringing with it a very particular set of connotations.

Apple had set out to offer a new kind of attitude towards software applications. Since the late 1990s, a phenomenon known as application 'bloat' had been increasingly evident. This is a tendency by software developers to keep on adding features to their applications, making each new version larger, slower and potentially more confusing to use. Apple had a very different conception of what an application for their new mobile devices should be: streamlined, functional, devoted to a particular task, free or inexpensive compared to traditional software, and far less elaborate to make and use (the fact that 'app' is also the first three letters of 'Apple' can't have hurt either).

Apple's decision effectively formalized the informal, making 'app' the term of choice for mobile applications – a shift in vocabulary that has helped mark the increasing intimacy between people and the digital devices in their lives. Similarly, Apple's iconic slogan 'there's an app for that' has helped breed the assumption that tailor-made assistance with almost any conceivable task need never be more than a few strokes of a touchscreen away.

Computer 'programs' and 'programming', today, are increasingly only for geeky specialists. For everyone else, there are simply apps. As far as most of the world is concerned, application software *is* software: eternally ready to apply the resources of ubiquitous computing to every aspect of living.

# 56.
# Fanboys and girls

The indulgence of passionate intensity is one of digital culture's defining features. Where once fellow-enthusiasts, nitpickers and aficionados tended to be divided by geography and united only via specialist magazines, associations and conventions, nobody today is more than a search term away from finding fellow travellers for even the most obscure obsessions.

It's a trend that has been reflected in the vocabulary of fandom. 'Fan' itself – in the sense of an enthusiast or follower – is a relatively young word, and the product of the late nineteenth century's potent combination of organized sports with mass print media. First used in the 1880s and 1890s to describe Americans devoted to following baseball, it probably emerged as a shortened form of the word 'fanatic', although it may also be related to the earlier habit of describing aficionados of a sport as having a 'fancy'.

More significant for our purposes, however, is a story told by – among others – Harry McCracken of the Technologizer blog, describing the appearance in 1973 of a home-made 'fanzine' (a word coined in the late 1940s by combining 'fan' and 'magazine') at a comic books convention in Chicago. The authors were writers and illustrators Jay Lynch and Glenn Bray, and the magazine was called *Fanboy*, a word Lynch had adapted from a piece of Florida slang from his childhood, 'funboy'.[69]

*Fanboy* magazine was a pencil-and-paper effort with an almost non-existent print run. Yet it gave the world a term for the particular obsessive species of fandom associated with comic collection and, soon, technology and gadgets.

Specifically, 'fanboy' has come to be used as a disparaging (although sometimes affectionate, especially if self-described) label for those with a particular loyalty to one brand, fictional world or other geek topic of choice. Search on Google for 'Apple fanboy' and you'll find around three-quarters of a million results, complete with a Wikipedia entry suggesting the synonym 'Apple evangelist'.

It wasn't until the second half of the 1990s that the notion of technology fanboys really took off. Since then, there has been no turning back, thanks in large part to the torrents of online content generated by those avidly interested in comics, science fiction and fantasy. As once 'cult' pursuits have become increasingly mainstream, the parallel term 'fangirl' has come into use, complete with a similar satirical edge. Meanwhile, a Japanese term coined in the early 1980s – *otaku*, derived ultimately from an honorific pronoun – has been adopted internationally as a description of extreme commitment to comics, video games or gadgets.

Today, to call someone who disagrees with you a 'fanboy' or 'fangirl' during an online debate is to level one of the commonest accusations in the digital realm: that they're blinded by obsession and have no sense of proportion. Equally for the increasingly huge numbers of self-identified fanboys and fangirls who populate both online forums and cult conventions like San Diego's legendary Comic-Con International, this may just be the proudest twenty-first-century title of all.

# 57.

# Welcome to the Guild

There's a little bit of the Middle Ages in some unexpected places online, and one of these is the realm of Massively Multiplayer Online Role-Playing Games (MMORPGs).

Consider the term 'guild'. In English, the word existed from the early thirteenth century and was originally spelled *yilde* – combining the Old English terms for an association of craftsmen or traders, *gegyld*, and for a payment, *gild*. It denoted a privileged association which one had to pay to join, and over time came to describe the powerful associations of traders who controlled exclusive rights to particular goods or markets in a region.

Although such guilds continued to exist in vestigial form in the twentieth century, the term had become largely used to invoke a medieval setting and connotations – something that made it the perfect fodder for the pseudo-medieval worlds of online fantasy gaming, as well as for their natural preoccupation with strict hierarchies and measures of expertise.

Gaming guilds are typically headed by a 'guild leader',

who wields quasi-dictatorial powers over his or her members along feudal lines. Below the guild leader exist a number of 'officers', sometimes with particular responsibilities: recruitment, the management of official guild websites or discussion forums, internal discipline. Below the officers come full 'members' of the guild, while below them are 'initiates' or probationary members.

As this kind of set-up suggests, gaming guilds can be every bit as concerned with pageantry and rigorous internal discipline as their antecedents. While some avowedly 'casual guilds' take a laissez-faire attitude, and 'social guilds' are mainly about hanging out with real or virtual friends online, serious 'raiding guilds' can entail stringent application processes, financial subscriptions, and minimum attendance pledges of ten hours a week or more.

Indeed, membership of the world's most elite gaming guilds – found perhaps above all in South Korea, where video-gaming is the country's second most watched television sport – can be a lucrative career for a fortunate few. More than fifty million guilds are thought to have risen and fallen worldwide during the last decade of online gaming – with the milieu even breeding its own hit web comedy series, *The Guild*, which over five seasons since 2007 has attracted over 80 million views around the world.[70]

Guilds aren't the only gaming groupings around. Often, some serious gaming collectives will also refer to themselves as 'clans': another word rich in historical and feudal associations (it's existed in English since the fifteenth century, originating in the Gaelic word *clann* meaning family or offspring).

For the most elite of all, though, it's the process of self-naming that's most important. Consider 'The Syndicate', a self-titled 'virtual community' founded in 1996 that today describes itself as 'the industry leaders in gaming excellence and the most successful virtual community in the history of online gaming'[71].

Complete with its own coat of arms and motto ('In friendship we conquer'), The Syndicate boasts an online 'charter' in the high medieval tradition, complete with mission statements, standards of conduct expected of members, and even an annual conference. For some modern netizens, playing games together is perhaps the world's most serious brand of fun.

# 58.

# Facepalms and *acting out*

Some of the central innovations of online communications lie in techniques for adding emotional and physical information to a conversation typed while seated at a computer. And some of these techniques involve a very particular layering of levels of self-awareness.

Consider the term 'facepalm' – a popular typed shorthand describing the act of slapping your own forehead with an open palm, to indicate wordless despair or frustration at someone else's words or actions. As a compounded word, 'facepalm' appears to date from the mid-2000s – often accompanied by a matching pop-cultural image, including one famous still shot of Captain Jean-Luc Picard from the 'Déjà Q' episode of *Star Trek: The Next Generation* – although the gesture itself is far older.[72]

What's interesting, however, is the precise meaning of writing such a word during an online exchange. When I type out the word 'facepalm', nobody actually thinks that I'm dropping my own head into my hand (even though I may be doing so). The agreed convention, rather, is that typing this neatly compressed term is an efficiently vivid way of suggesting – through a word – that I consider myself lost for words.

Especially when it forms a part of a larger sentence or description, facepalm is sometimes written within asterisks during an online chat. For example, if someone I was chatting to asked me 'So, is London a part of England?' I might reply 'I can't believe you're actually asking me that *facepalm*'. (With a nod to the typing conventions of online gaming, it's also increasingly common to write '/facepalm', as a forward slash is typically used to enter 'action' commands via an online game's chat interface.)

The typographic convention of placing a description of physical actions within asterisks was established early in the history of the internet as a standard form of chat room and forum etiquette, and one with richly inventive possibilities. Often, there's an element of humour or sardonic undercutting to the practice, or a desire to express something more complex than standard abbreviations for 'actions' like LOL or ROFL allow.

If I wish to be self-deprecating after introducing a complex idea, I might type something like '*crosses fingers, hopes no one will ask him to explain*'; if I feel embarrassed, it might be more along the lines of '*creeps into corner and closes eyes with shame*'. You'll notice that in each case the standard formulation involves the present tense and the third person singular, as if when I type an action inside asterisks I'm offering a running commentary on my own (hypothetical) actions from an external point of view.

It may sound deliberately obscure, or simply daft, but this ability to dramatize your own online communications is an important aspect of the full register of digital chat. Rather like characters offering whispered asides to their audience in a play, this kind of verbal mummery helps onscreen interactions achieve something of the liveliness of conversation in the flesh.

It also permits a range of physical reference that would be impossible if the actions described did actually have to be performed. Mock-suicide is, for example, one popular asterisked device — '*places gun against temple, pulls trigger*' — to which an approving response might be simply 'LOL' or '*gets out mop and bucket*'. Some online conversations are not for the faint-hearted.

# 59.
# Finding work as a mechanical Turk

One of the wonders of the late eighteenth century was a chess-playing automaton – that is, an entirely automated machine – most commonly known as 'the Turk'. Constructed in 1770 by the great inventor Wolfgang von Kempelen, it took the form of a figure wearing Turkish dress seated at a wooden cabinet with a chessboard on top. Opening the cabinet revealed some fiendishly complex clockwork, which apparently powered the Turk and allowed it to play a game of chess against any willing human: a challenge many of Europe's most celebrated figures accepted.

Almost all were defeated. Considered one of the great enigmas of its age, the Turk could even detect attempts at cheating by its opponents, beating Napoleon himself in 1809 despite the French emperor's efforts at foul play. It was, of course, a hoax – albeit of the most ingenious kind.

As was eventually discovered in the 1820s, the cabinet beneath the Turk was designed like a magician's box, concealing a diminutive human chess master who worked with the Turk's operator to create the illusion of a chess-playing machine – aided by magnetic counters underneath the chessboard, corresponding to the position of every piece, and a complex mechanism for moving the Turk's mechanical arm to pick up and place pieces on the board.[73]

The original Turk was destroyed by fire in 1854, but its name lives on today in one of the more ingenious forms of crowdsourcing to be found online: Amazon's very own 'mechanical Turk' service.

Launched in 2005, Amazon Mechanical Turk is a web-based service that invites users to register themselves as mechanical Turks: that is, to become the equivalent of the chess master crouched within the eighteenth-century mechanism. Other people then use the service and its legions of registered Turks to perform tasks that even today cannot automatically and efficiently be performed by machines: tasks such as correctly identifying the objects in a photograph, or potential sightings of missing persons.

Mechanical Turks working for Amazon earn a small amount of money for each 'Human Intelligence Task' or HIT performed, with tasks ranging from the cheap and simple ('View two images and determine whether they are the same kind of place, such as bathroom, forest or street' at 2 cents per answer) to the somewhat more involved and lucrative ('Check Webpage for Irrelevant Products' for a whole dollar). All of which is performed under the delightful slogan 'Artificial artificial intelligence'.[74]

The possibilities of mechanical Turks have inspired fictional as well as commercial imaginations, with one of the more striking ideas for their use featuring in science fiction author Cory Doctorow's 2010 novel *For the Win*. Set in a near-future of vastly complex online multiplayer games, one character, a boy called Leonard, earns money working as a Mechanical Turk for the (fictional) corporation Coca-Cola Games.

Rather than simply performing rote tasks, Leonard's role is effectively as a digital actor, taking over control of a non-player character (NPC) within the game whenever it encounters a situation beyond the parameters of a pre-prepared script: when, for example, another human player is behaving strangely or aggressively. It's a form of digital puppeteering not too distant from the practices of modern call centres – and an alarmingly convincing vision of a future in which you never quite know whether it's another human or a machine beyond your screen.[75]

# 60.

# Geocaching

New tools bring new possibilities for work and leisure alike, and one of the more curious among these is the sport of 'geocaching'. All you need to get involved is a GPS (that is, a Global Positioning System receiver, now found in almost every new smartphone) and information about the location of a 'cache': a container hidden at a particular location, typically containing a logbook.

With millions of active players, and caches hidden across the world, it's effectively history's largest treasure hunt. And it all began in May 2000, when the American government switched off the 'Selective Availability' technology it had hitherto been applying to GPS signals in order to ensure that only military equipment could use them to an accuracy of better than a few hundred feet.

With this, civilian GPS equipment suddenly became accurate to within a few tens of feet – and it took only a few days for computer engineer Dave Ulmer to announce to a newsgroup interested in satellite navigation that he had hidden a 'stash' at a particular set of co-ordinates.

Initially dubbed a 'GPS stash hunt', or 'gpsstashing' for short, some vigorous online discussion soon concluded that 'stash' had unfortunate negative connotations thanks to its link to drugs (indeed, the word entered English at the end of the eighteenth century directly from the realm of criminal slang) and that 'cache' was preferable (a word which arrived in English at around the same time as 'stash', coming from an identical French Canadian term used by trappers to denote where they hid their stores – a word itself derived from the French verb *cacher*, 'to conceal').

Credit for the term 'geocaching' is usually given to Matt Stum, a member of the original discussion group and pioneering geocacher. The prefix 'geo' itself derives from the ancient Greek word *ge*, 'earth', and occurs in numerous English words – as well as a few specialized items of geocaching vocabulary, such as 'geonick', which denotes the public nickname of a geocacher; and 'geoswag', denoting the goodies found concealed within some caches.[76]

While its use of technology is inherently new, geocaching itself fits into a long tradition of treasure hunts and games, the most significant of which is 'letterboxing'. Like many eccentric games, letterboxing dates back to Victorian England, and to walkers in Dartmoor in the west of England. According to an 1854 guidebook, a local tradition involved walkers leaving letters or postcards in a box on one of the trails across the moor, to be delivered at a later date by another walker who happened to come across the box.[77]

# Geocaching

Alongside geocaching, letterboxing itself still exists today in the United Kingdom and America, embracing a variety of different kinds of box. Some don't even physically exist, thanks to the young practice of concealing 'virtual letterboxes' within a website, consisting of an image of a letterbox whose location may be revealed by solving a number of online clues.

Both geocaching and letterboxing have their specialized vocabularies, including extensive acronyms for possible outcomes like 'Did Not Find' (DNF) and 'Found In Good Shape' (FIGS). One rather different word worth a mention is the common term for a non-geocacher: a 'muggle'. The word is borrowed directly from J. K. Rowling's Harry Potter books, in which it denoted a non-magic-using outsider. Being caught by such an outsider retrieving a geocache is known as being 'muggled', while a cache damaged or uncovered by a non-expert can also be described as 'muggled' — a textbook case of life and language imitating art.

# 61.
# The beasts of Baidu

With over half a billion people connected to the internet, China boasts the world's largest online population – and with this has come a sometimes brilliantly subversive digital culture of linguistic resistance to prevailing orthodoxies.

In 2009 the Chinese government added a process of keyword-based filtering to its arsenal of efforts to censor online content in keeping with its political and cultural values. Shortly after the introduction of this system, a series of mysterious articles began appearing on the much-used online Chinese encyclopaedia *Baidu Baike*, detailing the appearance and habits of four 'mythical creatures': the 'French-Croatian Squid', 'Grass Mud Horse', 'Chrysanthemum Silkworms' and 'Small Elegant Butterfly'.

Translated into English, they sound merely bizarre. In Chinese, however, speaking out loud the characters for each of these names results in a phrase very close to profanity – but, crucially, written using entirely different characters. So, for example, the phrase 'Grass Mud Horse' is spoken in Mandarin *cao ni ma* – while the phrase 'fuck your mother' uses exactly the same syllables, but pronounced with different tones, Mandarin Chinese having four different pitches that can make the same syllable mean completely different things. Similarly, saying 'French-Croatian Squid' in Mandarin produces the sounds *fa ke you* – a direct transliteration of the English phrase 'fuck you'.

Within days of their creation, these satirical 'beasts of Baidu' had gone viral. Among other things, they had soon spawned a spoof children's song on YouTube – complete with shots of cute alpaca-like beasts and a chorus of high-pitched voices – that attracted over a million viewers as it allegedly praised 'running grass mud horses' (a homophone for 'fuck your mother hard') and their rivalry with 'river crabs' (a complex pun on censorship, due to the similarity between the sounds of 'river crab' in Mandarin, *he xie*, and the official phrase for 'harmony', *he xie*, as preached by Hu Jintao and his state censors).[78]

Soon, the ranks of the legendary beasts of Baidu had swelled to ten, introducing citizens to characters including the *ji ba mao* or 'Lucky Journey Cat', whose name sounds extremely similar to the term for male pubic hair; and the *da fei ji* or 'Intelligent Fragrant Chicken', which resembles a slang phrase for masturbation.

All of which, thanks in large part to their delightedly daft offensiveness, have proved a highly effective way of both mocking and circumventing online censorship in one of the world's most digitally censorious nations. The 'grass mud horse' in particular has become an icon: a homophonic joke that's also an identifying marker for other subversive words and ideas, as well as an embarrassing reminder of the limitations of those seeking to control language and thought.

# 62.
# Snowclones

One of the internet's most powerful features as an influence on language is its ability to make the innovations of a minority instantly accessible to everyone else – so long as the innovation seems useful, amusing and/or important.

A case in point is an October 2003 entry to the estimable linguistics site Language Log, in which British-American linguist Geoffrey Pullum explained his belief that 'we need a name for . . . a multi-use, customizable, instantly recognizable, time-worn, quoted or misquoted phrase or sentence that can be used in an entirely open array of different jokey variants by lazy journalists and writers.' Pullum was asking his readers to come up with a word with which he hoped to skewer lazy thinking and language – and they duly obliged.[79]

Invitations to create new words are not exclusive to the digital age (although one of the most famous such stories, which holds that the word 'quiz' was introduced by an Irish theatre manager in the nineteenth century for a bet, is almost certainly apocryphal) but the business of doing so has grown immeasurably easier. So it was that, in January 2004, another language blogger, Glen Whitman, proposed on his site Agoraphilia that the word 'snowclone' be coined for the job, as an invocation of the thousands of writers using the hackneyed formulation 'If Eskimos have N words for snow . . .'. Pullum approved of the term, and duly declared it official.[80]

As Pullum also noted, one unique property of digital as opposed to merely written words is the precision with which we are able to record their transmission. So, he declared in a note to future lexicographers, 'since we have a record of the exact time at which Glen hit Send and transmitted the new term to me (the first person to read it), lexicographers are in luck here: they can date the coining of snowclone to . . . 22:56:57 (that's 3 seconds before 10:57 p.m.) on Thursday, January 15, 2004, in Northridge, California.'

A word used only by two people is one of little interest either to lexicographers or to anyone else. In the case of 'snowclone', however, the demand for this new term was evidently real enough, for it soon made its way off his blog and out into the wider world. Query Google about 'snowclones' today and you'll turn up over 100,000 results – including a 'snowclone database', representing almost five years' worth of dedicated digital spotting.

A particularly frequent offender here is the 'X is the new Y' formulation, as in 'green is the new black' – a phrase that has spawned generations of indolent offspring ('kite surfing is the new black'). But snowclone isn't the only word we can thank Language Log and its erudite operators for — another of the site's gifts to the world being the term 'eggcorn'. This time, the coiner was Geoffrey Pullum himself, in September 2003, the challenge being to coin a term for the habit of substituting one near-identical sounding word or phrase for another, as in the use of the words 'egg corn' to mean 'acorn' or — two cases that I frequently seem to encounter — the use of 'pendant' where 'penchant' is meant, or of 'gambit' where 'gamut' is meant.[81]

Eggcorn is worth still more hits than snowclone — and marks another gift to language conceived and propagated through the efforts of online word-watchers.

# 63.

# Typosquatting

Typographical errors have been with the world since the birth of movable type – one of history's most infamous examples being the publication in 1631 of the so-called Sinner's Bible which, thanks to the unfortunate omission of the word 'not' in Exodus 20:14, suggests that God's will is for His people to commit adultery.

The abbreviated term 'typo', though, has only been in use since the end of the nineteenth century, when the era of mass-typing first began thanks to the widespread use of typewriters. And it's only in the last couple of decades of mass online typing that an intriguing related phenomenon has sprung up: 'typosquatting'.

Typosquatting exists thanks to the simple fact that people often make typing errors when entering web addresses into their computers. This means that web addresses that use common misspellings of valuable, highly visited sites can themselves be worth considerable amounts of money in terms of advertising and traffic – or even as imitation versions of 'real' sites.

# Typosquatting

Typosquatting is technically a subset of the larger field of 'cybersquatting', in which legitimate sites, businesses, brands and individuals may find their names – or something very close to their names – being 'squatted' online by imitators looking for profit, or hoping to make mischief (the term 'squat' itself, meaning an illegal occupation of land, has been in use since the late eighteenth century).

Even one letter can make all the difference: visiting an address like 'wikapedia', for example, or 'twtter' (both these sites used to exist, but have since been shut down). Still more confusing is the fact that, if the owner of a correctly named website forgets to renew their web address, it can be possible for a squatter to leap into action and purchase the right to use that correctly spelled web address. Or this can be done in advance, buying up a web address ahead of the actual person or brand for whom it is named, allowing someone to 'squat' on what appears to be an official site.

At the more specialized malicious end of the spectrum, such phenomena as 'doppelganger domains' can be found: fraudulent websites identical to authentic ones except for the omission of a single dot between the domain and subdomain. For example, if a legitimate address included 'ca.ibm' then a fake one might use 'caibm' to fool users. These fake domains are often used in order to generate false emails which appear to come from a legitimate company, and can even involve such intricacies as automated routines for passing emails on to their intended recipients, and back again, having snooped on their contents – allowing a hidden third party to lurk in the middle of sensitive exchanges.

Finally, should you find yourself unsuspectingly lured to – or knowingly visiting – a malicious website, your web browsing experience may be marred by the practice known as 'mousetrapping'. This metaphor perfectly captures the rodent-in-a-cage feel of some web locations, where 'pop-up' windows open as fast as it's possible for you to close them, and efforts to leave the offending webpage result in simply being dumped back where they started. Typing one wrong letter can be a costly experience online (or a profitable one, depending on which side of the screen you're sitting).

# 64.

# Egosurfing and Googlegangers

Digital culture can be narcissistic, and one practice in particular embodies its most common form: 'egosurfing'. The word was coined in 1995, in a 'Jargon Watch' column for *Wired* magazine by Gareth Branwyn, and described the habit of looking up your own name in search engines — and quite possibly devoting many hours to surfing the results.[82]

Egosurfing has gone by several other names in its time, from vanity surfing to autogoogling, but it's the idea of 'ego' that probably best summarizes the tendency. Originally simply the Latin word for 'I', ego's modern sense of self-obsession is taken from the eighteenth-century metaphysical notion of the 'egoist' as someone who believes there isn't necessarily anything in the world apart from themselves — a usage that in due course gave rise to the common sense of an egoist as someone selfish.

If you are an egosurfing egoist, today, one potential obstacle to your self-delight is encountering a 'Google-ganger': a play on the word doppelganger, meaning someone else who shares your name and thus threatens to turn egosurfing into an investigation of someone else's life. If, for example, you happen to be called Tom Cruise but are not the famous actor of that name, you'll find it extremely difficult to egosurf effectively, thanks to the millions upon millions of results relating to the actor that any search for your own name will reveal.

Even if you don't have a globally famous namesake, Googlegangers can be a source of confusion in an age when other people are likely to look you up online as a primary source of information. Potential errors of conflation are compounded by the existence of sites and services that automatically compile information taken from search terms, and might thus mix together the details of four or five different people's lives if they shared the same name.

Finally, if you're looking for a broader indictment of the self-serving tendencies of the internet, the author and tech-sceptic Andrew Keen has been one recent advocate of 'digital narcissism' as a phrase for our times. The heart of this process, as Keen describes it, is an 'emptying out' of our inner selves by the constant pressure of self-broadcasting through social media: a modern echo of the legend of Narcissus himself who, having fallen in love with his beautiful reflection in a pool of water, starved to death. A cheerful image to bear in mind for those devoted to the eternal sunshine of their Facebook accounts.[83]

# 65.
# Infovores, digerati and hikikomori

What do you call the information-literate elite of the digital realm? Terms of celebration — and self-celebration — abound, among which one of the most widely used is 'digerati'. The word seems to have made its first appearance in print in January 1992, in a *New York Times* piece by John Markoff, referring to 'the computer digerati' as an expert class of the modern era.[84]

The word itself combines 'digital' with the idea of the 'literati' — a term in use since the early seventeenth century as a description of 'people of letters', the word being a plural form of the Latin *literatus* meaning 'lettered' (with perhaps a hint in contemporary usage, too, of the *illuminati* — 'the enlightened ones', originally used to describe a sixteenth-century religious sect).

If the digerati are the global super-elite of the information age, the increasingly numerous and enthusiastic consumers of their creations are sometimes labelled 'infovores': people who are hungry for information, and prone to glutting themselves on online knowledge.

A punning variant on words like herbivore and carnivore, the term 'infovore' was popularized by neuroscientists Irving Biederman and Edward Vessel, who in a 2006 paper on 'perceptual pleasure and the brain' explored humans' 'innate hunger for information' and argued that this 'quest for knowledge can never be sated' thanks to the brain's neurochemistry and the nature of the stimulation provided by learning.[85]

Biederman and Vessel's description is an excellent match for the web-browsing habits of millions of people – constantly drawn onwards by the urge to click on one more link, to follow up one more fact or search term – and so it's little wonder that 'infovore' has come to provide everything from the title of blogs and books (economist Tyler Cowen's *The Age of the Infovore* is perhaps the best-known example) to further scientific papers and research.

'Infovore' has also spawned an entertaining, if obscure, antonym: the 'ignotarian'. Coined via slightly spurious analogy to the term 'vegetarian' – a vegetarian being to a carnivore what an ignotarian is to an infovore – it implies someone who deliberately restricts their information consumption.

'Ignotarian' appears to have been coined in July 2006 on The Personal Genome blog, and – sadly – isn't in the widespread use it undoubtedly deserves.[86] Those looking to label the ugly underbelly of mass communications and information culture are probably better looking to Japan, and in particular to a term that has started to spread worldwide from the country of its birth: *hikikomori.*

*Hikikomori* literally means 'pulling inward', and was first used in Japan as early as the seventeenth century to describe people who pulled themselves away from society and lived as recluses. With the rise of computers and the internet, it has gained a whole new layer of meaning – not to mention fuelling an anxious debate that blames the phenomenon of people cutting themselves off from all forms of direct human contact on everything from bad parenting to video games.[87]

What's certain is that, for those driven to pull away from others, technology can be both a perfect means of self-isolation and a path towards re-engagement with the world. For infovores, as for most prolific consumers, it's all about finding a healthy digital diet.

# 66.
# Planking, owling and horsemanning

Using social media is often a kind of performance. It's an opportunity to show the world – or an extended circle of those you care about – just how interesting, amusing, attractive, intelligent, self-deprecating, athletic, or indeed pretty much anything else, you are.

Words remain the bread and butter of these self-presentations, but few things speak louder than a video or image – or have a potentially greater capacity to spread from mere hundreds of admiring viewers to millions. Enter, then, the phenomenon known as 'planking'.

If you want to go planking, all you need is yourself, a companion and a device equipped with a digital camera. Having gathered these and chosen your location, you then lie face down in a plank-like posture with your arms touching the sides of your body, and get your friend to take a photograph. You then put this photo online, and hope that the amusingly incongruous sight of someone lying face down in your chosen location will lead to the viral transmission of your image around the world.

## Planking, owling and horsemanning

As a word, 'planking' is evidently based on the idea of making your body deliberately stiff and plank-like: an analogy that holds all the way back to the Latin word *planca*, meaning a board or slab. The immediate source for the word 'plank' in English is the Old French *planche* – a term that remains in use today as a technical term for a technique in gymnastics where the body is held rigidly in parallel to the ground.

Like many online coinages, 'planking' can be deployed equally as a verb, adjective and verbal noun; also like many online activities, its profile was boosted in the most unfortunate of circumstances, thanks to a fatal incident in May 2011 when a young man died in Australia after planking on a high balcony.[88]

Online examples of what might be termed proto-planking exist from around 1994 and, not surprisingly, its venerable status as an activity has led to several more recent variations on the theme appearing. First documented in 2011 is a practice known as 'owling' where a person adopts a squat, owl-like posture and posts a photo online ('owl-planking' combines both, with one person adopting an owl-like posture perched on top of someone else planking).

Also born in 2011 were the practices of 'batting' or 'bat-manning' – where you hang upside down by your legs with your arms crossed – and most bizarrely of all, 'coning' or 'cone-ing'. This involves taking a video of yourself ordering an ice cream cone from a stall and then, rather

199

than simply accepting it, either getting out a second empty cone of your own and placing it on top of the ice cream you're being offered, or grabbing the ice cream topping off the cone and sticking it in your mouth. In either case, the crucial element is capturing the unsuspecting server's reaction on video, the 'point' being not the act itself, but the subsequent social media comments, likes, shares and reactions when it's uploaded.[89]

It's not all about the twenty-first century, though. As early as the 1920s, a fashion emerged among amateur photographers of staging photographs in which it looked as though someone's head had been detached from their body. The practice was termed 'horsemanning' – almost certainly after the headless horseman of Washington Irving's 1820 short story 'The Legend of Sleepy Hollow'.

Predictably enough, the growing digital fame of planking and its derivatives has today led both to a revival of horsemanning, and the scanning and republication online of many 1920s photos of the practice: a case of photo-posing trends coming full circle (and of people continuing to find the same kind of silliness extremely amusing).[90]

# 67.

# Unfriend, unfavourite (and friends)

'Friends' on Facebook are quite a different matter to friends in the rest of life. You may or may not count all your closest friends as 'friends' on Facebook, but you're almost certain to have 'friended' – or been 'friended' by – people for whom the conventional sense of that word is inappropriate.

My own Facebook 'friends list' includes many close members of my own family, some of my closest friends from life and work, and a whole host of professional and tangential contacts, as well as people whom I've never even met in person. This is fairly typical. The fact that a social network with around a billion registered accounts chose 'to friend' as its principal verb of interconnection has not so much shifted the older sense of the word as created an entirely new one – drawing attention in the process to both the social network's aspirations, and the gulf between its rhetoric and actuality.

The word 'friend' itself comes almost directly from the Old English word *freond*, itself derived from the verb *freogan*, meaning to love or bestow favour upon. The idea of 'friending' as well as 'befriending' has been used as a verb for over half a millennium – but it wasn't until the public advent of Facebook in 2005 that its contemporary sense arrived.

The use of the warmly emotional term friend, rather than the somewhat colder social networking notion of 'contacts', is central to Facebook's ethos. It has also created a particularly grating linguistic contortion in the form of the verb to 'unfriend' someone, in the sense of removing them from your Facebook contacts list. It's both a marvellously economical formulation and a reflection of the binary principles on which social networking is based, where all it takes is a click of a button completely to negate an action or status.

To be 'unfriended' is, moreover, to be instantaneously cut off from someone else without even knowing it has happened. While the social network is quick to announce incoming 'friend requests' to its users, the process of being unfriended happens silently and without notification of any kind: a fact that, inevitably, has bred a website called 'Unfriend finder' expressly designed to help you discover who no longer wants to be your friend.[91]

Perhaps a more accurately descriptive term for social media contacts can be found in Twitter's popularizing of 'followers' and the idea of 'following' someone as a description. Again, there's an echo of a physical original in the description (and the word's origins are, like those of 'friend,'

# Unfriend, unfavourite (and friends)

Old English, from *folgian*, 'to accompany or pursue') but the metaphorical sense is also ancient. What is new, however, is the inevitable inversion of 'unfollow' – together with the notion of a 'follower count' as an increasingly important online status symbol, allowing people to put a precise figure on the level of public interest in their own words and actions. Young verbs like 'to unfavourite' and 'to un-like' follow in these binary footsteps, giving the lie along the way to the fuzzy feeling of comfort that words like 'favourite' are presumably intended to evoke.

With each follower being, at least theoretically, an actual person who has actively chosen to follow somebody else, the question 'how many followers do you have?' is fast becoming a universal index of impact for public personalities and brands alike – with the title of world's most-followed Twitter personality hotly competed. At the time of writing, in mid-2012, Lady Gaga led the world with over 27 million followers – around ten million more than the world's most-followed politician, Barack Obama; and twenty million ahead of the world's most-followed geek, Microsoft founder Bill Gates. Never has it been easier precisely to measure how much less the world is interested in you than in somebody else.

# 68.

# Sneakernets and meatspace

If someone suggests you use the 'sneakernet' to transfer a file, don't reach to turn on your computer. With its deliberate echo of 'ethernet', the term is a pun at the expense of technology, and refers to the practice of physically transferring information from one place to another. In this case it's by putting on a pair of sneakers and walking with digital information held on your person within some kind of physical storage medium: a disk, memory stick, and so on.

Sneakernets, sometimes, can be considerably faster as a method of data transfer than the internet itself. A courier delivering a package of high-capacity disks between two locations in a city, for example, generally constitutes a far higher rate of data transfer than if the files had been sent electronically.

The function of terms like sneakernet is both to amuse and to make people think twice about the limitations of the systems they're using. In terms of computer security, a highly secure network may nevertheless be breached by someone using the sneakernet to infiltrate it: that is, by someone physically copying information from a computer onto a disk, and then walking out with that disk in their pocket.

A breach of this kind is exactly what US serviceman Bradley Manning allegedly perpetrated in 2010 when hundreds of thousands of military files were copied onto a CD (labelled 'Lady Gaga') and then handed over to the Wikileaks organization for publication online. Similarly, those wishing to avoid detection may use sneakernets to transfer files from their computers to other locations for uploading in order to evade detection, the greatest online security guarantee lying in not being attached to the internet at all.

'Sneakernet' is just one of a number of words used to poke implicit fun at the ways in which the digital realm can make what was once ordinary seem strange. Another is 'meatspace' – otherwise known as physical reality. The word is a punning echo of 'cyberspace' and has its origins in William Gibson's 1984 novel *Neuromancer*, in which the protagonist, an elite hacker, dismissively refers to his physical body as 'meat' and the offline realm as the 'meat world' (although the precise formulation 'meatspace' didn't itself arrive until the early 1990s).[92]

This technique of making the once-familiar seem strange can also be seen in phrases like the 'dead-tree press' as a name for traditional, print newspapers. What's explicit here, however, is that what was once considered ordinary is now not only strange, but also increasingly ridiculous and outdated in a digital age – and likely (at least according to the coiners and users of the phrase) not to exist for much longer. In this sense, terms like 'dead-tree press' are also a kind of prophecy: a prediction that the objects they describe will one day feel as ridiculous to everyone as the phrase itself aims to sound.

# 69.
# Going viral

The word 'virus' is unashamedly unpleasant in its etymology. Arriving in English in the late fourteenth century,
it came directly from the Latin term *virus* for a poison or
unpleasant, slimy liquid. As a modern medical term, 'virus'
was coined in the 1890s by the Dutch microbiologist
Martinus Beijerinck, who used it to describe an infectious
agent far smaller than bacteria whose presence he had
detected.

Beijerinck believed that viruses were, like their Latin
namesake, some kind of contagious liquid. By the 1930s,
however, it had become clear that they were in fact largely
made out of protein, and subsequently that they consisted
of genetic information inside a protein coating. It's also
debatable whether viruses do or don't constitute life as such:
able to reproduce only when inside the cells of other living
organisms, they infect all known living organisms, yet can
have no existence of their own outside of other life.

It was this analogy of parasitic infection that led to
the label 'virus' being applied to the notion of malicious,

self-replicating computer programs – first by science fiction author David Gerrold in his 1972 novel *When Harlie Was One*,[93] and then to describe the similar programs that began to infect personal computer systems in the early 1980s. Like its biological counterpart, a malicious program is technically only classified as a virus if it needs human actions to act as 'hosts' to continue its spread – unlike computer 'worms', which can spread themselves automatically.

Today, however, it's the notion of 'viral spread' via the internet that's the most important digital use of the term: a phenomenon that dates back to the early 1990s, and the recognition that words, ideas and images could all exhibit virus-like behaviour under certain conditions, thanks to the entirely conscious and willing actions of their human audience.

'Virality' is one of the stranger properties a digital creation can be said to possess. It's not so much about what something is, inherently, as about how people react to it. Something 'goes viral' online when a sufficient number of people link to it and spread the word – a process that can look very much like the fast-forwarded spread of a virulent infection across the global population. Yet this spread is inherently impossible to predict, or engineer.

As I explored in the earlier entry on memes, famous viral phenomena range in their ridiculousness: from videos of a toddler biting someone's finger to captioned photos of cute cats, political protest videos against African warlords, and misleadingly titled links to Rick Astley music videos.

Since the late 1990s, this has led to such increasingly lucrative and widespread practices as 'viral marketing', where firms use any and all of the digital resources at their disposal to try to ensure the viral spread of brand and product awareness. Such campaigns often co-opt the strange, subversive and crowd-friendly tendencies of 'authentic' viral phenomena in order to seduce and delight their audiences – and to persuade them to engage in the all-important acts of digital sharing upon which every would-be-viral message ultimately relies.

As with much else online, every viral act – no matter how vast the statistics involved – ultimately relies upon a conscious, individual human choice; a willing decision to play host to part of the torrents of information tearing across the world.

# 70.
# Dyson spheres and digital dreams

One curious language category is terms for things that don't exist yet, but might. Consider the concept of 'Dyson spheres'. Named after the great physicist Freeman Dyson, who first hypothesized them in 1960, these spheres would consist of a network of structures orbiting a star that would effectively form a shell around it, capturing almost all of its energy for use as a power source capable of fuelling technologies more powerful than any currently conceivable.

Dyson was engaging in a radical kind of speculation, but one that he argued was linked to present concerns by a firm chain of reasoning. Harnessing the power of suns would, he theorized, eventually be an essential energy resource for any sufficiently advanced civilization. It thus follows that in our present attempts to identify any extraterrestrial life in the universe, searching for the tell-tale infrared energy signatures of any such constructions might be one way of locating advanced alien civilizations.[94]

The very notion of Dyson spheres has itself driven

further technological speculations, including the so-called 'matrioshka brain' – the brainchild of the late futurologist Robert Bradbury, who named his hypothetical creation after Russian matrioshka dolls, which stack neatly inside each other in a set of ever-decreasing size.[95]

The matrioshka brain would be a supercomputer of almost unimaginable potential power, consisting of several specially designed Dyson spheres around a star, each one drawing power from the sun's energy to drive billions of tiny computers, and then passing on some of that energy to the next shell-like layer of the brain. Such a machine could, at least hypothetically, exist without violating the known laws of physics – and would possess a level of computing power able to achieve near-miraculous results.

Among the possible applications of such a vastly powerful computer is the operation of a completely convincing 'simulated reality' – a concept long known in science fiction, and connected to a number of the most imaginatively compelling of hypothetical technologies. One of these is the notion of human 'uploading': that is, copying the complete data representing a human life and memory into a vast super-computer, able either to store that data or to run it within a simulated reality, offering human beings a kind of immortality.

The possibility that we may already be living in such a simulation is one that both authors and philosophers have debated in some form for many centuries. One particularly appealing innovation in this field is the American author Greg Bear's fictional coinage of the 'Taylor algorithms': algorithms that allow a being trapped within a simulated reality to determine the true nature of their world.[96]

Testing the boundaries of the thinkable as well as the merely feasible is one of language's great delights – and technology is, today, perhaps its most fertile arena. As the author Arthur C. Clarke once put it, 'the only way of discovering the limits of the possible is to venture a little way past them into the impossible' – a journey for which only the right words are required.

# 71.

# Welcome to teh interwebs

One of the more intriguing aspects of a typed medium consisting of billions of words is the use of deliberate misspellings to signify a particular tone.

Take the jocular formulation 'interwebs', often combined with 'teh' to give 'teh interwebs'. Typically, it's used by those who think of themselves as digitally experienced to poke fun at the supposed naivety of those who aren't – who might, for example, consider calling the internet the 'interweb', a conflation of 'internet' and 'World Wide Web'.

As a comical term, 'interweb' is a child of the mid-2000s; as a science fiction term, speculatively describing the connection of multiple internets across different planets, it dates back at least to a 1994 episode of the television series *Babylon 5*, an origin that may well have helped convince the unwitting that it did in fact describe something real.[97]

The history of the plural form, though, adds another layer to the joke, invoking a specific incident during the

US presidential campaign in 2000 when George W. Bush described during a public debate the idea that 'we can have filters on Internets' as a measure against pornography and violence.[98]

Saying 'the internets' rapidly became a knowing way of poking fun at those seeking to discuss technology without necessarily understanding it; combining it with online culture's fondness for whimsical typos like 'teh' augmented the gag further, as does the occasional habit of (mis)spelling 'interwebs' as 'interwebz' or even 'intarwebz'.

Misspellings and substitutions for comic effect are nothing new, of course – in 1775 the playwright Richard Sheridan's character Mrs Malaprop gave her name (a 'malapropism') to the unintentional misuse of one similar-sounding word for another.

Online, though, rampant and systematic misspelling is above all a visual gag, rooted in the specific mechanics of a keyboard – something that draws extensively on the trends and codes outlined in this book's earlier entry on 'leetspeak'. In everyday online usage perhaps the closest historical analogue to teh internets and friends is writers' use of 'eye dialect'. This phrase was coined in 1925 by the American professor of English George Philip Krapp to describe the use of unusual spellings in fiction to convey the fact that a character was speaking strangely, but without making any attempt actually to write out their dialect phonetically.[99]

An eye dialect, in other words, is a form of writing

aimed at the readers' eye rather than ear. It signals difference and idiosyncrasy in an efficient, comprehensible
way — such as in Mark Twain's use of the spelling 'wuz'
instead of 'was' for some of his characters — without
demanding a perhaps hard-to-read transcription of the
precise sound of speech itself.

The internet is, in this sense, the apogee of eye dialects,
with its plethora of terms and typos aimed entirely at the
eye, — and echoing not speech, but the conversational
deployment of a clacking keyboard.

# 72.

# On good authority

In thirteenth-century English, an *auctorite* was a text that could settle an argument by virtue of its impeccable standing. The highest *auctorite*, naturally, was the Bible, followed by the greatest works of the classical world, and words written in the name of the king or church. The word came to English via Old French, but originated in the Latin term *auctoritas*, meaning 'prestige' or 'influence'. *Auctoritas* was derived from the Latin word *auctor*, meaning a 'leader' or 'founder' – a word itself based on the verb *augere*, 'to increase or augment', and which directly led to the English word 'author'.

Etymologically, then, authority is rooted in the idea of tracing something back to its origins: to the person or text who originally caused something to thrive and increase. In terms of power structures, this came in medieval English to be equated with those canonical texts from which all truth flowed, and then with those wielding power in the name of this truth – that is, the king and the church, and all those patronized by them.

# On good authority

Over time, as the feudal hierarchy was challenged by more enlightened and rational thinking, ideas of both authority and authorship came increasingly to privilege those who were forging new expertise and creating new texts rather than simply demanding deference to ancient words and hierarchies.

Perhaps the most significant shift in its meaning for several centuries, however, came with the publication in 1998 of a paper called 'The Anatomy of a Large-Scale Hypertextual Web Search Engine' by two graduates in the computer science department at Stanford University: Sergey Brin and Lawrence Page.[100]

Brin and Page outlined 'a prototype of a large-scale search engine . . . designed to crawl and index the Web efficiently and produce much more satisfying search results than existing systems'. To achieve this, they offered an algorithm – punningly termed PageRank – which calculated for every page on the web 'an objective measure of its citation importance', based on the principle that more authoritative pieces of research tended both to be cited more frequently by subsequent research, and to cite each other more frequently.

Brin and Page had, in other words, come up with a ranking system inspired by academic research that was able automatically to calculate the 'authority' of online resources – and it was this new principle of discernment that underpinned much of the early success of their search engine, Google, when it was founded later that year.

The principles of Google's search algorithm are far more complex than a simple measuring of links; and it did not invent the idea of online authority as such. What its huge success has helped demonstrate, however, is a fundamental change in the nature of what it means to offer an authoritative answer to a query in the digital age. Because it is possible, today, to provide algorithmic and statistical responses based upon mass observation to almost all of those questions that once upon a time simply involved locating an authoritative source. Moreover, the very act of seeking out authority has become digitally mediated.

Ask Google a question, and the answers will come back ranked in order of putative usefulness and reliability from one to however many million responses it has located. Search for a word and the number of results will offer an index of its global adoption. Where once authority may have seemed an impervious, god-given (or book-granted) right, it looks today more like something both earned and measured by the ways in which the world does (or doesn't) choose to heed what is being said.

# 73.
# A world of hardware

'In October 1953', begins the first sentence of the introduction to technologist Paul Niquette's online book *The Software Age*, 'I coined the word "software."' A young programmer at UCLA – which boasted one of the just sixteen digital computers that existed in the world at that time – Niquette had, he claims, come up with the term as a jokey way of distinguishing the programs people like him were creating from the physical 'hardware' of digital computers themselves.[101]

'Software' didn't actually appear in print until five years later – courtesy of American statistician John Tukey – but enjoyed informal usage among early computer scientists for some years before then. 'Hardware', meanwhile, has been with us in some form for far longer; and it is to the form of the word hardware that we ultimately owe numerous elements of modern computing terminology.

In the sense of equipment made of metal, hardware has existed in English since the fifteenth century, deriving ultimately from the Old English words *heard*, 'hard, firm, solid'; and *waru*, 'a valuable object'.

The notion of 'hardware stores' had become common in America by the start of the nineteenth century, and it made obvious sense to extend the term to describe the machine components that went into building the earliest computers – a usage first recorded in 1947, and common by the end of the 1950s. Hence Niquette's gag, and the coining of a term for something new to the world: an engineered, functional tool consisting entirely of information.

As well as a huge number of phrases and specialisms, one effect of the introduction of 'software' as a punning companion term to hardware has been a new etymological life for the word 'ware' itself. This is most obvious in the formation of further words to describe types of program: 'spyware' (software intended surreptitiously to track users' actions), 'malware' (a generic term for malicious software, intended to cause harm or somehow exploit users), 'freeware' (free software), 'shareware' (free trials of software intended to be shared with others), and so on.

Beyond this, 'ware' has also become a new word in its own right among some computer users, and hackers in particular. Mutated into the plural form 'warez' – the 'z' for 's' substitution being a typical swap in hacker subculture – the word has come to describe pirated versions of various kinds of software, usually distributed illegally or via shadowy, informal services.

'The warez scene' – sometimes shortened simply to 'The

# A world of hardware

Scene' — is thus the semi-official term for the international community of those who specialize in distributing pirated software, as well as numerous other kinds of media content: films, television shows, music, video games. It's a scene that has been a part of the internet since the early 1970s — well before the world wide web came along — but that new technologies and systems have fed. It's also a far cry from innocent punning among the world's few computer programmers half a century earlier; although not, perhaps, from that original Old English sense of something precious, and thus eminently worth stealing.

# 74.

# Darknets, mysterious onions and bitcoins

Underworlds have their own dialects, and the internet is no exception. Underneath the searchable, visible, day-to-day services used by most people lies a realm of 'darknets' – hidden networks and methods of information exchange used by those who, for a variety of reasons, don't want anyone else to find out what they're doing.

Apparently coined in a 2002 Microsoft research paper, 'The Darknet and the Future of Content Distribution', the term's vividness owes a debt to science fiction. What it describes, though, is an eminently practical business.

One key to unlocking the darknet is the use of a specialized program that grants its users anonymity, the most common of which is 'The onion router' system, usually simply abbreviated to 'Tor'. While the name may sound arbitrary, it is in fact descriptive: developed in the late 1990s, 'onion routing' is a system that encrypts messages several times and then sends them through multiple points on the network known as routers. The system works by allowing each router to remove one layer of encryption

222

before passing on the message to the next router: a process analogous to peeling an onion, and which means that the content of messages themselves remains hidden at all points along its journey.[102]

Once Tor client software has been downloaded, users can theoretically participate in various darknet activities – which may vary from the evasion of censorship and anonymous contact between sources in countries with highly restrictive regimes to the grim businesses of illegal drugs, pornography and so on.

Unlike the conventional internet, navigating darknets isn't as simple as searching for what you want. Instead, resources like the 'Hidden Wiki' tend to be used. As opposed to what's sometimes referred to as the 'clearnet' – that is, the ordinary internet, accessible through ordinary browsers – the Hidden Wiki can only be accessed through Tor services, ensuring anonymity, and contains a linked listing of darknet resources.

One of the most controversial recent developments among darknet sites is known as Silk Road. Named in honour of the great east–west global network of trade routes along which silk was once exported from China, it was launched in early 2011 as an anonymous marketplace for more-or-less anything that anyone might wish to buy or sell online.

Also accessible only through Tor services, Silk Road is perhaps most intriguing because of its exclusive use of 'bitcoins' as currency for transactions. In monetary terms, bitcoins are sometimes called a 'crypto-currency', and constitute one of the most radical experiments to date in constructing an electronic cash system entirely based on cryptographic technology – creating a seemingly foolproof, decentralized system able to regulate itself without the need for any government or central authority oversight.[103]

Initially released at the start of 2009, bitcoins were the brainchild of programmer Satoshi Nakamoto (almost certainly not his real name), and are 'mined' at a predetermined rate by all users running the bitcoin software, with an eventual limit to their total circulation set at 21 million. Around eight million had been created by the start of 2012, with a real-world monetary value of close to fifty million dollars at current exchange rates. Truly a term and a concept fit for the digital century.

# 75.

# Nets, webs and capital letters

The word 'internet' — or simply the 'net' to its friends — is with us largely (although not entirely) thanks to the United States Department of Defense.

It all began in 1969, with the launch of the Advanced Research Projects Agency Network, or ARPANET, which linked together four separate networks of small computers at four different sites in America. It was, in effect, a network of networks, connected to each other thanks to a new protocol for data exchange known as 'packet switching'.

Thus arose the notion of calling this set-up an 'inter-network' — a term that seems to have appeared first in its contracted form, 'internet', in a 1974 document outlining the specifications of the system written by Vint Cerf, one of that handful of American computer scientists honoured today with the title 'the fathers of the internet'.[104]

An interesting footnote to the word itself is the question of whether the initial 'i' of 'internet' should be capitalized or not. The argument for capitalization rests on the point that there is only one internet, representing the interconnected mass of every sub-network in the world operating according to the Internet Protocol (IP) that Cerf helped draw up.

Back in the 1990s, the capitalized form of Internet tended to be used by most official publications, from newspapers to researchers. Today, though, the balance has shifted the other way. The argument against capitalization is that, essentially, the word is now so common that it's absurd to treat it like a proper noun, especially when most people think of themselves not so much as users of a monolithic Internet as users of distinct internet-based services, from email to social networks.

Alongside the net, perhaps the most universal term in digital use today is the 'web' – a shortened form of 'world wide web'. A far younger beast than the internet, both the web itself and the word describing it exist thanks to just a handful of people employed in the late 1980s at CERN, the European Institute for Nuclear Research.

Chief among these was Tim Berners-Lee, whose March 1989 proposals for 'a large hypertext database with typed links' provided the basis of the system he and his collaborator Robert Cailliau subsequently launched publicly in August 1991. As Berners-Lee explains in his book *Weaving the Web* it was nearly called something very different.

Alternative names under consideration included The Information Mine (rejected because the acronym TIM spelt out Berners-Lee's first name) and the Mine Of

Information (rejected because its acronym, MOI, was also the French word for 'me'). The Information Mesh, another potential name, was turned down because it sounded too much like 'mess'. And so the phrase World Wide Web won out, thanks to its stress on what Berners-Lee described as his brainchild's 'decentralized form allowing anything to link to anything'.[105]

In 'www', too, it offered a unique and memorable acronym — whose nine syllables make it the most time-consuming three-letter spoken combination it's possible to create in English, as well as one of the few acronyms that actually takes longer to pronounce than the phrase it is 'short' for.

# 76.
# Praying to Isidore and tweeting the Pope

Isidore of Seville was born around the year 560 in Cartagena, Spain. He rose to become archbishop of Seville and one of the greatest writers and historians of his age, as well as a noted religious leader who oversaw the conversion of many of the Visigoths ruling the once-Roman province of Iberia to Christianity.

Isidore died in 636, and was canonized as a saint just under a millennium later, by Pope Clement VIII in 1598. For our purposes, however, the most significant date in his posthumous career came in 2006 when the Vatican officially named him the patron saint of computer technicians, computer users, computers, schoolchildren, students, and of the internet itself.

Isidore's patronage was inspired above all by his greatest written work, the *Etymologiae*. This is an encyclopaedia that across twenty volumes attempted to preserve in summary form all of the classical knowledge Christians felt was worth preserving: a kind of medieval meta-database of the world's knowledge, some 1,500 years ahead of Wikipedia.

## Praying to Isidore and tweeting the Pope

The official nomination by the Vatican followed in the footsteps of considerable previous online enthusiasm. Some Spanish Christians had already been invoking Isidore for several years as an official 'cyber protector', while as early as 2000 the the Order of Saint Isidore of Seville was founded as an online organization devoted to 'Promoting the ideals of Christian chivalry through the medium of the Internet'.[106]

Among other resources, the Order offers a prayer to be spoken before using the internet, whose advice it's difficult to disagree with: 'During our journeys through the Internet / We will direct our hands and eyes / Only to that which is pleasing to Thee / And treat with charity and patience / All those souls whom we encounter.'

The Vatican itself has enthusiastically embraced many aspects of technology over the last decade, with a slick website, mobile site, apps and even an official papal Twitter account: the rather oddly named @Pope2YouVatican, which links up with an official 'Pope to You' website – a phrase that seems unlikely to breed a global internet phenomenon.[107]

As the secretary of the Pontifical Council for Social Communications pithily puts it, 'many of the key Gospel ideas are readily rendered in just 140 characters'[108] – a practice that surely begs the coining of a snappy phrase to match the twentieth-century birth of 'televangelism'. 'Internet evangelism' doesn't quite do it, not least because the term has already been taken up by those enthusiastic simply about technology rather than God, while the notion of 'cyberchurches' doesn't go far enough. All suggestions will no doubt be most welcome on @Pope2YouVatican.

# 77.

# QWERTY and Dvorak

For centuries, the mechanisms used to play instruments like church organs across Europe were known only as 'manuals' (a word in use from the start of the fifteenth century, meaning something operated by hand – from the Latin *manualis*, meaning 'of the hand').

Then, around 1820, a new term arrived on the scene: the 'keyboard'. A more literally descriptive term than 'manual', within a few decades its usage had begun to be extended to other machinery – and when the world's first commercial typewriter was put on sale in 1870, 'keyboard' became the preferred term for describing the layout of its letters.

It wasn't until the success of a typewriter model released by American manufacturers Remington in 1878 (the inventively named Remington No. 2, their second typewriter model), however, that the most famous keyboard layout in the modern world made its way into the mainstream: the QWERTY arrangement.

QWERTY – named after the first six letters on the

keyboard, reading from the top left – was designed specifically to allow users to type fast in English without jamming the metal arms on which letters were mounted, something that tended to happen when adjacent letters were repeatedly used. The QWERTY layout thus ensures that most common letter combinations within English words are split across the keyboard. Similarly, the diagonal arrangement of the letters – which are offset beneath one another, rather than appearing in a straightforward grid – was designed to make clashes between levers less likely.

In modified form, QWERTY remains the standard for most keyboards using the Roman alphabet around the world today. Even though mechanical levers have long vanished from the process of typing, there can be few computer-users alive today for whom the letter arrangement QWERTYUIOP isn't instantly familiar.

QWERTY's dominance comes partly thanks to historical momentum, but it does have at least one serious rival, albeit with a far smaller pool of users. This is the Dvorak keyboard layout – named not after its particular arrangement of letters, but in honour of its co-inventor, Dr August Dvorak, who together with his brother patented his brainchild in 1936.

Dvorak's keyboard – known properly as the Dvorak Simplified Keyboard (DSK) – was based on the principle of attempting to reduce to a minimum the distance travelled by someone's fingers when typing in English. With a top row of letters beginning with punctuation before reading PYFGCRL, to most computer users' eyes it looks jarringly alien compared to the QWERTY approach, although its exponents argue that it can both increase typing speed and reduce errors.

In any case, if you wish to try it for yourself, most modern computers include the option to switch from QWERTY to Dvorak – while some keyboards will even let you remove their letters and numbers and replace them wherever you see fit (although you'll also need to recon-figure your computer if this isn't utterly to bewilder anyone using it after you).

# 78.

# Apples are the only fruit

How did what is now the world's most valuable company come to be named after a fruit? According to Apple's co-founder, the late Steve Jobs – as related to Walter Isaacson in his 2011 biography – 'he [Jobs] said he was "on one of my fruitarian diets." He said he had just come back from an apple farm, and thought the name sounded "fun, spirited and not intimidating."'[109]

This was in Cupertino, California, in April 1976, when the young Steve Jobs and his friend Steve Wozniak released what they called the Apple I computer. Rumour has it that Jobs also admired the music company the Beatles had set up – Apple Corps, founded in 1968 in London – and that he considered apples a beautiful fruit for a logo (the original Apple I manual also featured an illustration of Isaac Newton reading underneath an apple tree – an intellectual precedent for the kind of innovation Apple hoped to embody).

Some eight years later, fruit were in mind once again when Apple was in the process of releasing its prototype home computer for non-expert consumers. The project had been started in the late 1970s by Apple employee and inter-face expert Jeff Raskin, who originally wanted to name it after his favourite variety of apple, the McIntosh Red: a green and red fruit originating from trees cultivated in the early nineteenth century by a Scottish-Canadian farmer named John McIntosh (the original tree, planted around 1800, survived until 1910, and its site is marked today with a headstone).

The name 'McIntosh', however, proved too close to an existing brand of audio equipment, and so the spelling was changed to 'Macintosh'. Finally released in 1984, the Apple Mac – as it soon became affectionately known – revolutionized the home computing industry, as well as initiating a series of computers that has continued to be developed and produced to this day.

In the twenty-first century, Apple has perhaps become best known – linguistically at least – for its use of the prefix 'i' to denote its products. This began just before the millennium, with the 1998 release of the 'iMac', the first major new product produced by the company after Steve Jobs's return as CEO in 1997 after an absence of a dozen years.

With the subsequent triumphs of iPods, iPhones, iTunes and iPads, that 'i' has become an international badge of computing cool. Yet, if Jobs had gone with his first choice of name, all this would have been very different. As Apple's advertising creative director Ken Segall tells the story, Jobs unveiled the firm's new computer internally under the provisional name the 'MacMan', and claimed to hate the name 'iMac' when it was presented to him as an alternative. Fortunately for all involved, Segall insisted on presenting exactly the same name to him again a week later; and this time, deciding he no longer actively hated it, Jobs allowed the iMac to be born.[110]

# 79.

# Eponymous branding

Digital vocabulary is rife with brand names, as befits a space in which every single action means using several services devised by other people. The most famous have become everyday words in their own right – 'Googling', together with the increasingly common verb 'to Facebook' – while even those that haven't taken on this kind of familiarity still constitute some of the world's most internationally recognized words.

Companies creating digital services can often seem as weightlessly abstracted as their products themselves: indeed, there are few human creations that more effectively alienate us from the circumstances of their production than online tools or the machines we use to access them. Dig beneath the surface of many of these names, and you'll find particular geographies and histories waiting to be uncovered.

As well as being the name of a vast communications corporation, for example, Nokia is also the name of the small town in south-west Finland where that company's second pulp mill was built in 1868 three years after its founding (in 2008 the company's last operations in the town were moved away further south). Similarly – albeit more recently – American software giant Adobe Systems was named after the Adobe Creek which ran near the house of its co-founder, John Warnock, in California.

The names of people and places run throughout the world of digital brands, often referring to still-living people, given the industry's youth. The open-source operating system Linux, initially released in 1991, takes its name from the first name of its creator, Linus Torvalds, who at the time was a student in Helsinki. Torvalds thought using his own name for the software was egotistical; but his co-worker Ari Lemmke felt that Torvalds's proposed name, Freax, was less memorable, and took matters into his own hands.[111]

Digital names don't always have any link to local geography, of course, a case in point being the online retailing giant Amazon. Its founder, Jeff Bezos, was looking for a name that both came near the start of the dictionary and that embodied something he hoped would become the biggest of its kind. The Amazon river met both these needs, having the greatest flow of any river system – a perfect metaphor for Bezos's desire to see all the world's online commerce flow through his site.

Finally, some geographies are buried deeper than others within famous names. In 1995, an online auction house website called AuctionWeb was founded in California by programmer Pierre Omidyar. Two years later, Omidyar changed the company's name and web address, originally aiming to bring it in line with his consulting company, Echo Bay Technology Group. But his first choice of domain name, echobay.com, was already being used by a Californian mining company working in Echo Bay, Nevada. Omidyar thus went with his second choice, ebay.com – and the rest is history.[112]

# 80.
# Mice, mouses and grafacons

The computer mouse has one of the most straightforward etymologies of any digital device. Consisting of a track-ball housed in a conveniently hand-sized plastic casing, and linked to the computer by a wire, its appearance resembled that of an archetypical mouse – and so this affectionate name for a novel method of controlling a computer soon became its official label.

The first recorded reference to a computer 'mouse' came in 1965, in a joint publication by W. K. English, D. C. Englebart and Bonnie Huddart, entitled 'Computer-Aided Display Control'. Although English contributed to the development of the device, however, and Huddart was involved in the team that tested it, the credit for its conception must go to Engelbart.[113]

Engelbart's invention came to him in 1963 at the Stanford Research Institute, when he and English were going through some of Engelbart's old notes on possible pointing devices for use in computer displays – one of which entailed using small wheels attached to an object to track the movement of someone's hand over a desktop. Based on Engelbart's notes and sketches, English carved a working prototype out of wood, its centre hollowed out to allow the wheels to be placed underneath.

The mouse had been born — but it didn't formally gain its name, or acceptance as a standard item of computer equipment, until a contract from NASA led Englebart and his team formally to investigate the usefulness of different interface devices in their 1965 paper. The mouse proved a clear winner — and in doing so earned both its permanent place in their future work, and its moniker in English's published paper.

'Within comfortable reach of the users' right hand', reads the first reference, 'is a device called the "mouse," which we developed for evaluation.' It concludes that 'subjects found that it was satisfying to use and caused little fatigue'. Other intriguing devices tested at the same time included a light pen, joystick, 'Grafacon' (a tablet consisting of a grid of wires embedded in a flat surface, across which a stylus could be moved) and an intriguingly labelled 'knee control' which, although only at an early stage of development, was noted as 'very promising', not least because it left both hands free.

Today, the mouse has become a near-ubiquitous computing tool, thanks in part to Apple's seminal 1984 Macintosh computer, which popularized 'point and click' as a central, elegant model for everyday computing. In one area at least, however, profound linguistic uncertainty remains: what exactly is the plural of 'a computer mouse'? The *OED* generously suggests that both 'mouses' and 'mice' are acceptable – but, to my ears at least, 'mouses' is ugly bordering on unacceptable. Unless, of course, there's an urgent need to avoid ambiguity, as in this unlikely sentence: 'The mice nested next to the mouses I keep in my desk.' Then, perhaps, it's permissible.

# 81.

# Meh

While new words are often devoted to emotional extremes, there's a special place in many hearts for the supremely useful three letters of the exclamation 'meh', which express an almost infinitely flexible contemporary species of anti-climax or indifference.

This flexibility is embodied in the uses to which 'meh' can be put. In its basic exclamatory form – most often found during online chat, emails, or other digital communications – it communicates something along the lines of 'okay, whatever, it doesn't bother me (but you should consider me determinedly unimpressed by the whole affair).'

As an adjective, it takes on a more ineffable flavour: 'it was all very meh'. It can also be modified to convey the same sense for something more concrete, as in 'the meh-ness of last night's dinner'. You can even use it as a noun, if you feel the need: 'I stand by my meh: the evening was forgettable.'

Some authorities note that 'meh' advanced in popular culture thanks to the patronage of US cartoon series *The*

*Simpsons*, where it seems to have featured for the first time in 1995 as a generic uninterested retort[114] – a kind of low-energy parallel to Homer Simpson's iconic cry of frustration, 'D'oh!'

In the pre-Simpsons era, one theory is that meh originated from the Yiddish term *mnyeh*, an exclamation with a similarly disdainful meaning.[115] It's thanks to the dominance of online typing and text messages, though, that 'meh' has really come into its own as an all-purpose, one-word response. The verbal version of a nonplussed shrug, it today boasts close to a hundred million Google hits to its name, not to mention a place in the *Collins English Dictionary*.[116]

As the British author and critic Sam Leith neatly observed in a 2008 column for the *Daily Telegraph*, reflecting on the word's ascent towards canonical status, meh 'denotes mild boredom rather than the extreme form. That subtlety of meaning is its value. I can't think of an equivalent.'[117] In this, it's a perfect emblem of the variations of register so important to online discourse, where typed language must match all the emotions necessary for casual conversation – and all the fine gradations of indifference.

# 82.

# Learn Olbanian!

English isn't the only language with elaborate online variations; and one language with an especially highly
developed alternative online form is Russian. Since the
late 1990s, an elaborate system known as Padonkaffsky
jargon has been a feature of the Russian-language internet,
fuelled initially by intellectuals and academics, but today
an integral part of Russian pop culture as a whole.

The name Padonkaffsky itself signifies the jargon used
by so-called *padonki* – an online counter-cultural movement
who named themselves after the plural of the Russian word
*padonok*, a slang term for people of low status. The *padonki*
first forged their jargon on marginal websites known for
sometimes obscene content, with Padonkaffsky jargon relying
on phonetic misspellings of Russian words combined with
complex puns and tongue-in-cheek cultural references; techniques that constitute a truly international recipe for
converting your native tongue into net-worthy jargon.

The nickname 'Olbanian' for Padonkaffsky is itself a
pun on the Albanian language (which has nothing to do

with the jargon), and came to fame in a 2004 incident where an English-speaking user on the website LiveJournal complained about not being able to read some text, which was written in Padonkaffsky-influenced Russian.

Having forcefully asserted that everyone ought to be conducting discussions on LiveJournal in English – and been mischievously told that the language he couldn't read was Albanian – the hapless user found himself bombarded with messages advising him to 'learn Albanian!' as a satirical way of highlighting English-speakers' inability to understand anything other than their own language, let alone jargon variations on other languages using non-Roman alphabets.[118]

'Learn Albanian!' rapidly went viral as a trend, becoming – in its Padonkaffsky form 'Olbanian' – a rallying cry for Russian language and pop culture online, as well as a humorous stock response to anyone using incorrect grammar or talking nonsense. Soon, 'Olbanian' had become a widely recognized alternative label for Padonkaffsky itself.

Sometimes written as !Olbanian! perhaps the most famous word in its vocabulary is *preved*, a deliberate misspelling of the word *privet*, an informal greeting. *Preved* became notorious as a Russian internet meme thanks to its appearance in a series of comically captioned pictures, beginning with a drawing of a bear interrupting a couple having sex, and culminating in sufficient public awareness for 'preved' to be used as the main caption on a Russian advertisement for *Newsweek* magazine, complete with an excitable figure mimicking the bear's posture.[119]

While it may sound bizarre, the 'bear surprise' image, as it's usually known, deserves a special mention. The image itself is a childlike watercolour painting by the American painter John Lurie, and was originally captioned with the word 'surprise'. Through the wonders of Olbanian, this became *preved*, while the bear himself earned the name 'Medved'. Since its first appearance in 2006, *Preved Medved* has earned an astonishing level of recognition – further evidence, if it were needed, of the weird transnational power of web-fuelled language games.[120]

# 83.
# Booting and rebooting

Why do we 'boot up' computers? The term dates back to the mid-1970s, and was first used as a shortened form of the term 'bootstrapping'. 'Bootstrapping' had itself been in use since at least 1953 by computer scientists, featuring that year in a document by Werner Buchholz on 'The system design of the IBM type 701 computer', which described 'a technique sometimes called the "bootstrap technique"' constituting a form of 'self-loading'.[121]

Starting a computer involves, from a programming perspective, making sure it executes a number of instructions one after another, beginning with the most essential processes and then, after these have been started, moving on to higher functions. 'Bootstrapping' proved a convenient metaphor for this thanks to the well-established English metaphor of 'pulling yourself up by your bootstraps' – that is, grabbing hold of the leather straps affixed to the back of gentlemen's boots and somehow using these to pull yourself up.

By the time it was adopted for computer science this

metaphorical notion of 'bootstrapping' had come to describe someone managing to get themselves going, under their own efforts. Before the twentieth century, however, people took rather more literally the impossibility of actually managing to lift yourself up via your own bootstraps, and the phrase was usually deployed to describe someone undertaking an impossible task.

A pleasing folk etymology for this original sense of the phrase is the 1785 narrative *The Surprising Adventures of Baron Munchausen*, whose eponymous protagonist boasts of achieving impossible feats – including managing to lift himself out of a swamp by pulling on his own hair. There's no actual reference to bootstraps within these adventures, though, and it's thus more likely that the idiom began in America at the start of the nineteenth century as a simple reference to an absurd impossibility.

Booting has come a long way since then, and can be found today in a variety of technological forms: from 'quick boots' and 'multi boots' to 'rebooting' a computer when starting it up doesn't work the first time. Then, of course, comes the ultimate act of computer science expertise: performing a 'hard reboot', otherwise known as turning a machine off and on again at the plug.

# 84.

# Cookie monsters

Many modern websites prominently announce to users that, in order to use them, you must have 'cookies' enabled in your web browser. Although it sounds like a request for baked goods, computer cookies are tinier and far more numerous. They are small pieces of data used by websites to store information on a user's computer, in order to provide a working record of previous actions, and thus to serve various functions such as authenticating a user, keeping track of purchases or inputs, and saving information about them between visits.

Computer cookies are properly known as HTTP (Hypertext Transfer Protocol, the basic data foundation of the web) cookies, and were named after a computing concept called the 'magic cookie' principle. Magic cookies embodied the idea of using small packets of data as tokens of exchange between a user and computer program: the data token would be used to represent a user's past actions, and could effectively be 'handed in' to the same or a different program in future. The data itself would be encoded, making it effectively invisible to the programs in question until it was actually handed over – an everyday kind of informational 'magic'.

As to the word 'cookie' itself, one of the more popular theories about its origin points to the Xerox Corporation in the early 1970s, where some employees first came up with the idea of storing small chunks of information about users' actions that could be exchanged between different computers and programs. Inspired by the popular television programme the *Andy Williams Show*, in which a bear often followed around the protagonist demanding a gift of cookies, they decided to give their little exchangeable crumbs of data the same name.[122]

In the sense of the sweet biscuit Andy Williams's bear was desperately seeking, 'cookie' entered American English language at the very start of the eighteenth century, coming from the Dutch word *koekje* meaning a 'little cake'. It wasn't until 1994, however, that cookies first found their way onto the world wide web, courtesy of the programmer Lou Montulli, who realized that the practice of using magic cookies to communicate information securely between programs had enormous potential for enhancing the functionality of websites – and duly coined the term in its online sense.[123]

Web cookies today come in a variety of flavours: from 'secure' cookies, that are permanently encrypted when being transmitted, to 'third-party' cookies, which are generated by online services distinct from the website you happen to be visiting.

There are also potentially troublesome cookie types,

including 'tracking' cookies (sometimes known as 'persistent' cookies), which can be used to generate a long-term record of someone's use of a website; or 'zombie' cookies, which are automatically recreated even if a user attempts to delete or remove them.

Despite the cute name, cookies can be a controversial part of the hidden infrastructures of online information, as anyone living in the EU will have noticed thanks to its recent 'cookie disclaimer' laws, making it compulsory for all websites to 'obtain consent to store a cookie on a user or subscriber's device'. Even when it sounds like a tasty snack, information is often an indigestible business.[124]

# 85.
# Going digitally native

What do you call someone born into the age of the internet and web – that is, from around 1990 onwards? Inspired by Douglas Coupland's 1991 novel *Generation X*, some plump for 'Generation Z' – the 'Y' generation having been born in the 1980s. Perhaps the most popular term out there, however, is one with a nicely pseudo-anthropological ring: a 'digital native'.

Digital natives, as the name implies, are the first generation born into a world where the internet, world wide web and their associated digital technologies were not only to be taken for granted, but were also older than the people using them.

The word 'native' itself originates from the Latin *nativus* meaning something innate or produced by birth, and it's interesting to speculate what it means to possess once-radical technologies as a birthright. What's also interesting, though, is the sense in which words like 'native' implicitly cast digital technologies as a landscape, or even a land, into which it is possible to be born as a kind of citizen: technologies that are not so much mere tools as a space within which living itself occurs.

This sentiment is equally evident in a term complementary to digital natives that describes the generations preceding them: 'digital immigrants'. The metaphor of technology as a physical environment is the same, yet its implications are rather different. For it is the elder generation – the immigrants – who are the interlopers and who, implicitly, must adapt their ways if they are to survive.

Although similar metaphors had been in use since the mid-1990s, both terms were popularized by the same person: the author Marc Prensky in a 2001 article entitled 'Digital Natives, Digital Immigrants', which framed questions of new technology in the context of education. 'Today's students,' Prensky argued, 'are no longer the people our educational system was designed to teach.' For Prensky, those who have grown up in the new digital era 'think and process information fundamentally differently from their predecessors'. Small wonder, then, that their digital 'immigrant' instructors 'speak an outdated language (that of the pre-digital age) [and] are struggling to teach a population that speaks an entirely new language'.[125]

What's most interesting about Prenksy's thesis is not whether it's an entirely accurate summary of intergenerational relationships with technology, but its enormous success in propagating its terms. The idea of technological progress as a yawning divide between generations has struck a powerful chord, together with the notion that it's the older 'immigrants' who are today struggling to keep up, and must either learn new customs or risk irrelevance.

Another intriguing aspect of the native/immigrant argument is its emphasis on language itself; and on the ways in which language change around digital technologies can involve the fundamental reconfiguring of old habits and associations. The simple act of watching a video, for example, exists in two different conceptual worlds today: one reflecting the traditional twentieth-century model of sitting on a settee in front of a television; another reflecting the rising twenty-first-century norm of streaming a video in one window on a computer via Wi-Fi while sharing your responses to it live via social media.

The same could be said for most other forms of media consumption – from what it means to 'read' or 'look up' something, to the associations surrounding 'listening to music', or the notion of 'media consumption' in the first place. Immigration pains may prove an unfortunately accurate metaphor.

# 86.

# Netiquette and netizens

The word 'etiquette' came to English around the middle of the eighteenth century courtesy of French, where *étiquette* described the properly prescribed behaviour for a particular situation – a term derived ultimately from the Old French *estiquette*, meaning a label or ticket.

'Netiquette', naturally enough, means 'internet etiquette', and was punningly coined in the early 1990s at a time when the recent advent of the world wide web had suddenly begun to open up the internet to a host of new, inexperienced users.

It's a phrase that may seem to be a contradiction in terms, given the libertarian bent of much online culture. Yet the original aspirations at its heart were paternalistic in the best geeky sense: an attempt by the internet's expert early adopters to offer practical advice (and even wisdom) to the flood of new users arriving online.

One of the earliest references to the word 'netiquette' itself was in a 1995 memo from the Intel Corporation offering 'a minimum set of guidelines for Network

## Netiquette and netizens

Etiquette (Netiquette)' formulated as part of the work of the Internet Engineering Taskforce, a standards organization founded in 1986 – originally as a US government-funded body – to promote and develop online standards of both the technological and ideological kind.[126]

The 1995 memo distilled many of the conventions of online responsibility that had developed over the previous decade of internet use by, primarily, academic and military institutions – a remarkable number of which still ring true today. 'If you are forwarding or re-posting a message you've received,' it observes of email, 'do not change the wording. If the message was a personal message to you and you are re-posting to a group, you should ask permission first.' And, most importantly of all, 'Never send chain letters via electronic mail.'

Similarly, the guiding principle of civilized interactions it offers is an enduring one – 'Be conservative in what you send and liberal in what you receive' – while it also offers sensible lessons on the act of typing: 'Use mixed case. UPPER CASE LOOKS AS IF YOU'RE SHOUTING.'

Like many social phenomena, the meaning of netiquette has shifted with the medium's rising popularity from being an attempt at prescription to something largely descriptive: an articulation of online social norms, and a warning against certain digital taboos. In this sense, a broader idea of responsible online actions has emerged that is sometimes summarized by description of internet users as 'netizens': citizens of a digital space, within which a degree of civility is owed. For the more idealistic, there's even been a 'Declaration of the Independence of Cyberspace', courtesy of one John Perry Barlow in 1996, hailing 'a world that all may enter without privilege or prejudice'.

Despite its libertarian tendencies, most of the net's social taboos today focus on various kinds of dishonesty or deception: the failure to acknowledge others as the originators of ideas, or to link to their sites or words (the practice of politely citing your sources is known as 'hat-tipping', a delightful metaphorical invocation of a once-literal social nicety). Similarly, although the online realm can be a free-for-all, it's also a place in which it can prove extraordinarily hard to get away with prominent false claims, given the ways in which determined digital detectives can sniff out data trails.

This isn't to say that anonymity online doesn't breed all kinds of excess and exploitation. But there are countervailing forces – and determined disruption may, to its surprise, be met with an equally determined enforcement of integrity, albeit of a very particular kind.

# 87.

# The names of the games

By some measures, Mario – the chubby moustachioed hero of Nintendo's best-selling series of video games – is the world's most recognizable fictional icon. Far fewer people, however, know how it is that a Japanese company previously best known for manufacturing playing cards created this beloved personage in 1981, and decided to make him a New York-based Italian plumber.

The 1981 video game that introduced Mario to the world was called Donkey Kong, itself the result of an unlikely naming process whose history varies depending on who you ask. The game saw players controlling a small, jumping man who was attempting to climb up a hazardous building site in order to rescue a woman from an irate giant ape: the 'kong' of the title, thanks to Japanese slang derived from the title of the film *King Kong*. The 'donkey' half, according to game designer Shigeru Miyamoto, was simply intended to invoke stupidity or goofiness – although some still claim that it was actually a miscommunication of the word 'monkey'.[127]

In any case, when the game appeared it was not its giant malicious monkey who would command the most attention, but its diminutive leaping hero, whom players had to guide through multiple screens of increasingly challenging obstacles – a then revolutionary development in the design of video games. The player-character was nicknamed 'jump man', thanks to his ability to jump over the obstacles tossed at him, and Miyamoto initially planned to make his official title 'Mr Video'.

According to gaming legend, it was subsequently decided to give the character a more personable name based on his resemblance to Nintendo's warehouse landlord in America, one Mario Segale. The story has never been officially confirmed, either by Nintendo or Segale. The name, however, has proved a triumph – together with Mario's unlikely status as an Italian-American plumber, a role only created for him as part of his second video-game outing, 1983's 'Mario Brothers', in order to fit in with his onscreen habit of running through giant pipes.[128]

Alongside Mario, perhaps the greatest early icon of arcade gaming is a still more mysterious character, Pac-man – if yellow circles with mouths can actually be called 'characters'. First released in 1980 by another Japanese firm, Namco, playing the game means manoeuvring your yellow hero around a maze, gobbling up dots while avoiding evil ghosts.

Thanks to this consumption-centred task and the game's crude sound effects, the character was initially named 'pakku-man', after the Japanese phrase *paku-paku taberu*, describing the sound of someone repeatedly opening and closing their mouth. Upon its release in Japan, this name was simplified to 'Puck Man'. And there it might have remained, had the game not been bought for distribution in America, leading Namco to recognize that having 'puck' in its title might prove too great a temptation for teenagers who spotted what happened when the initial 'p' was replaced with a slightly different letter.[129]

# 88.
# Flash crowds, mobs, and the slashdot effect

It all begins, once again, with science fiction. In 1973 author Larry Niven published the novella *Flash Crowd*, which imagined a world in which the invention of affordable mass teleportation devices meant that, within seconds of any newsworthy event taking place, huge numbers of people would almost instantly arrive on the scene.

A futuristic vision of technologically enhanced rubbernecking (itself a delightful slang word that emerged at the very end of the nineteenth century in America to describe someone whose neck seemed to be made of rubber, so intent were they on listening in on others' conversations), Niven's title was subsequently taken up to describe a virtual version of the same effect: the sudden arrival of huge numbers of people on one webpage, usually via a link from an enormously popular site.

One of the first major generators of online flash crowds was the technology site Slashdot – so much so, in fact, that the term 'Slashdot effect' is often used to describe a sudden, overwhelming influx of online traffic thanks to

## Flash crowds, mobs, and the slashdot effect

a popular link. Slashdot itself was founded in 1997, offering a collection of user-submitted links to technology news stories and articles complete with lively comment threads. With over five million monthly users, receiving a prominent link from Slashdot can cause smaller sites to be overwhelmed – a phenomenon sometimes simply termed 'being slashdotted'.

Inevitably, other popular sites have attracted their own variants of this verb. Crashing under the weight of visits sent your way by popular news aggregation site The Drudge Report is known as being 'Drudged', while links from the humorous aggregation site Fark can lead to your site being 'Farked'. More inventively, a deluge of web traffic coming from the popular American politics blog Instapundit is known as an 'Instalanche'.

Returning to the original notion of flash crowds, cheap mass teleportation remains a long way distant technologically, but what we do possess is cheap mass organizational tools – something that has helped create the analogous phenomenon of 'flash mobs'. A flash mob is a crowd of people who have used new media to assemble rapidly in a public place, usually in order to perform an apparently spontaneous act of co-ordinated protest, art or advertising.

The world's first successful flash mob was organized in June 2003 by one Bill Wasik, an editor at *Harper's Magazine*, who arranged for participants in four locations in New York to gather in groups of over one hundred people and await instructions. When they came, these involved the seemingly-spontaneous performance of absurdist actions (precisely fifteen seconds of mass applause from the mezzanine level of the Grand Hyatt hotel; two hundred people simultaneously turning up to buy a single rug in Macy's department store). As Wasik puts it, the experiment used mass communications to create 'an inexplicable mob of people . . . for ten minutes or less'.[130]

If Wasik's experiment was intended as satire, it's a fine irony that flash mobs today are perhaps above all associated with filmed examples used by brands for self-consciously trendy marketing campaigns. More interestingly, though, there's also the growing creation of what's known as 'smart mobs': large crowds of people whose massed behaviour is intelligently co-ordinated by the use of smartphones and other digital devices.

The creation of smart mobs has helped fundamentally alter the dynamics of events such as political protests, with plugged-in members of the crowd able to access the kind of information and capacities for co-ordinated action previously only possessed by trained military and law-enforcement units. Intelligent group behaviour doesn't have to be political – and examples have included co-ordinated crowd performances at concerts and the mass use of cameras to record and broadcast an event – but there are powerful political implications in the fact that most members of modern crowds are now anything but mob-like in their capacities for action, reaction and communication.

# 89.
# Godwin's law

Online discussions aren't famed for their reasonableness or politeness, but there is a kind of elegance to be found in the way that certain ideas and analogies recur throughout the medium. Of all these, Godwin's law describes one of the oldest and most persistent: the likening of those who disagree with you to Hitler and Nazi Germany.

Godwin's Rule of Nazi Analogies, as it's sometimes more formally known, was first formulated by American author and attorney Mike Godwin in 1990 to describe a trend he had observed on Usenet discussion forums: 'As a Usenet discussion grows longer, the probability of a comparison involving Nazis or Hitler approaches one.' In other words, any online discussion that continues for long enough will inexorably end up invoking Nazis or Hitler.[131]

Godwin's point was that, once a discussion reached this state, it had self-evidently ceased to be a proper discussion, and therefore the only sensible option was to abandon it (one unintended corollary of which being that someone

wanting a discussion to be abandoned might intentionally invoke Hitler).

Over time, Godwin's law has come to stand as a handy
indictment of any form of argument online that resorts
to labelling one's opponents as the worst thing imaginable:
an accusation anticipated in 1951 by the philosopher Leo
Strauss when he coined the phrase *reductio ad Hitlerum*
as a variation on the *reductio ad absurdum* (the Latin for
'reduction to absurdity') form of false argument.

At the heart of Godwin's observation lies not an assault
on false arguments, but rather an attempt to highlight
how thoughtless comparisons can help trivialize even
appalling things. It's a trivialization that's fully in evidence in a host of other contemporary idioms, some of
which manage to be sufficiently charming or absurd to
defy their subject matter.

'Kitlers', for example, are photographs of cute cats that
look like Hitler, while one of the most astonishingly successful viral videos of recent years involves various titles
beginning 'Hitler finds out that . . .' followed by a section
of the German film *Downfall*, in which a German-
language scene imagining Hitler's deluded fury during
his last hours in his bunker is provided with comic sub-
titles (in a typical version, he might be commenting on
the results of a football match, or discovering that his
favourite surfboard won't be ready for a forthcoming trip).

Perhaps the final word should go to Godwin himself,
reflecting in *Wired* magazine on the success of what he

termed his 'counter-meme' – that is, his attempt to oppose a common, unreflecting mode of argument with a neatly embodied counter-argument. 'If it's possible to generate effective counter-memes,' he asked, 'is there any moral imperative to do so? . . . Do we have an obligation to improve our informational environment? Our social environment?'[132] In an information-saturated age, it's perhaps one of the more important questions we can ask.

# 90.

# From Beta to Alpha to Golden Master

Writing computer software is an iterative business. You build a basic version, test it, try to fix what isn't working, add further features, and repeat – a process that is likely to continue for as long as your software continues to be used, given the impossibility of ironing out every potential error experienced by every potential user. As an old programmers' joke has it: What's the similarity between computer programming and sex? One mistake and you have to support it for the rest of your life.

Naturally, a specialized vocabulary has emerged around this process. Although much of it consists simply of jargon or oblique acronyms, it has also gifted the world a number of more universal labels for the stage of a product's development: from its 'pre-alpha' and 'alpha' beginnings on through 'beta' to the eventual 'golden master' edition.

All of these terms are thought to have their roots in the hardware development process used by technology firm IBM as early as the 1950s, when stages labelled A, B and C marked a product's development and testing. As the company and its peers moved from hardware to software development, the 'pre-alpha' stage became increasingly important: that is, the research and initial design of a product.[133]

An 'alpha' product traditionally marks the first stage at which testing can begin, with 'beta' marking the phase at which all the main features have been added, but many bugs still need to be identified and addressed. The process of 'beta testing' a product typically involves both a 'closed' phase (when it is tested internally by the company producing it, or by a select outside group) and an 'open' phase (when the public and ordinary users try a working version).

Finally, a 'golden master' edition results from this process: a term named after the practice of producing the final versions of physical media such as vinyl records using a 'master' copy literally made out of gold, so that it could be used as an unreactive template from which all subsequent copies would be created.

Today, it's increasingly common for online software to be publicly released in its beta phase, and to remain in beta for many months or even years. Sometimes referred to as 'perpetual beta', this embodies the fact that the testing and fine-tuning of a digital product can represent a potentially endless undertaking.

While frustrating for some, this is hardly surprising, given that a product developed by a few dozen programmers but used online by many millions of people is likely to demand an immense amount of maintenance, updating and iteration. Indeed, the very notion of finished versions of some software is fast becoming outdated, with one of the worst fates that can befall a program being the status of 'abandonware': something for which no further updates or support will ever be produced.

# 91.

# Mothers and daughters, masters and slaves

Personifying our machines is often irresistible, and even some of our most technical language reflects this. Consider something as basic as the plugs and wires coming out of a computer, and the shape of the slots they fit into. For physiological reasons that shouldn't need to be spelled out, any connection that has protruding elements – such as the prongs on a plug – is called a 'male' class of connector; while anything featuring slots for these protrusions to fit into is called a 'female' connector.

In its excellent (if occasionally disconcerting) entry on 'gender of connectors and fasteners', Wikipedia notes that the gender analogy is almost universal in fields entailing the manufacture of connecting parts: from mechanical design to construction toys, plumbing, ductwork, electronics and computing. There are even such things as 'her-maphroditic' connectors – which include both male and female elements – and 'gender mender' devices (also known as 'homosexual adaptors') which allow male-on-male or female-on-female connections.[134]

The mechanical use of the term 'male' dates all the way back to the 1660s. Considerably more morally weighted than the language of gender, however, is the language of power relations that describes a connection between multiple electronic devices (or processes) where one has complete control over the others. Known as the master/slave protocol, it most typically applies to computer hard disks, where the 'master' disk dictates the functioning of other 'slave disks'.

In widespread use since the 1990s, the term pairing made headlines in November 2003 when the County of Los Angeles's purchasing department asked manufacturers supplying it not to use these terms, thanks in part to a discrimination complaint from an employee earlier that year − a stance that hasn't much dented the terms' use elsewhere.[135]

Altogether more charming are some of the descriptions of the actual components within a computer. First coined in the mid-1950s, a computer's 'motherboard' is the large framework onto which all its other major components are attached. Combined with the literal 'board' that formed the basis for early circuits, 'mother' seems to have been used by analogy with nineteenth-century terms such as 'mother-ship'. Pleasingly, together with this term arrived the description of smaller circuit boards attached to the motherboard as 'baby boards' − or, later, 'daughterboards' (sometimes also known as 'piggybackboards').[136]

Why circuit boards should be exclusively female is not

easily explained — although the engineering analogy to the naming of ships as 'she' may have played its part. Along, perhaps, with the affection in which they were held among the largely male ranks of early computer scientists.

# 92.

# Bit rot

Bit rot sounds, at first hearing, like a paradoxical proposition: how can digital information decay, let alone 'rot' (a fine Old English word, originally *rotian*, 'to putrefy')?

The idea behind bit rot is that the individual units of electrical charge constituting digital bits can disperse over time. More generally, though, it has become a term used to describe the degeneration of physical media on which data is stored: from CDs and DVDs to older formats, such as magnetic tape and disks.

Physical damage isn't the only challenge facing those who wish to preserve digital information, or even the most severe. Far more problematic is the basic fact that any digital information is useless if you don't have access to the software and hardware designed to read it, and to convert that raw data back into words, images, code, or whatever it is that they embodied in the first place.

Already, some of the world's earliest computer data, generated by institutions like NASA in the 1950s and 1960s, has become almost impossible to access thanks to the destruction of the physical machinery needed to read it – and that's before the physical decay of the storage medium itself comes into play.

As well as bit rot, a slightly more enigmatic phenomenon known as 'software rot' can also plague older computers and their programs. Software rot – also sometimes known as software entropy – is not a physical process, but rather reflects the ways in which software gradually tends to work less and less effectively as it is transferred between computer systems, and as the machines on which it is operating become more distant in time from the period in which the software itself was created.

There are, for example, many programs from the early days of computing which it is simply impossible to run on any modern computer, as their instructions are no longer fully comprehensible by, or compatible with, modern computers and the programming languages running on them.

Software that has ceased to be usable on modern machines is sometimes referred to as 'legacy' software, which may have to be 'refactored' if it is to operate on a modern system – or be run through a specially designed 'emulator' program, which effectively recreates a virtual version of an older system within a modern computer.

Arcane though it may sound, issues of emulation and data preservation are some of the central problems facing anyone wishing to preserve the young history of computing in a usable form. Ultimately, digital data and programs are all a kind of language; and unless systems remain that can read and speak all the many languages of their past, much of modern history risks becoming literally incomprehensible.

# 93.

# Non-printing characters

There's precious little romance in the phrase 'non-printing characters' – and yet behind it lurks something of a noble pedigree. Non-printing characters are a series of symbols that it's possible to display within a word-processed document which won't show if you print out that document, but which indicate where particular pieces of formatting have been applied, such as paragraph and page breaks, spaces, tabs, and even simple spaces.

Among these largely ignored (and usually invisible) elements of the typed realm, one in particular stands out: the pilcrow, or ¶. Also sometimes known as a paragraph mark, it does indeed mark those points in a document at which a paragraph break has been inserted. Its history, though, is far older even than movable type.

The pilcrow symbol itself developed from the medieval practice of marking sections within a written document with a letter 'C', standing for the Latin term *capitulum* meaning a 'chapter'. Over time, and thanks to scribal haste, the 'C' was elided with two vertical strokes, a marking that originally indicated an instruction from the scribe writing out a text to other scribes who might be adding to or ornamenting it.

Unlikely though it may seem, the word 'pilcrow' itself is simply an extremely corrupt version of the term 'paragraph', with the same origin in the Greek term *paragraphos*, a written stroke that signified a shift in the sense of a text (from *para-*, meaning 'beside', and *graphein*, 'to write'). The word first appeared in English in 1440 in the form 'pylcrafte', having been − it's thought − mangled via old French and scribal transmission.[137]

Pilcrows, today, covertly adorn almost every word-processed document created − and can be made visible should an author wish, together with a host of other non-printing characters that includes line breaks (depicted as a bent arrow pointing to the left), page breaks (a dotted line with the words 'Page Break' in its middle) and even marks to indicate typed spaces (a tiny dot in the middle of the line for each one), sometimes referred to as 'white-space' marks.

All of these are sometimes known as 'control characters', reflecting the ways in which they determine the appearance of digital text while themselves remaining unseen. They are a kind of programming language for text, complete with a historical pedigree as ancient as the Greek practice of marking the margins of manuscripts with explanatory symbols − including the very first examples of the *paragraphos* itself.

# 94.

# Wise web wizards

In fifteenth-century England, the word 'wizard' emerged as, initially, a description of someone wise – a 'wise-ard', from the Middle English word *wys*, meaning wise. By the middle of the sixteenth century, however, the word had taken on some of the occult connotations it retains today; the Middle Ages were a time when reason and mysticism were seen as aligned rather than opposed, and it was natural that a wise person would also possess insights into occult fields.

Until the twentieth century, wizards remained venerable (and largely male) personages, although the term came to refer almost exclusively to fantastical figures involved with magical forces rather than actual, wise people in everyday life. In the 1920s, the term also took on a pop cultural second life as a general exclamation of approval ('wizard!') – a revival paralleled by its rehabilitation as a mainstream written term through the growing body of 'high fantasy' literature, and in particular thanks to the publication in 1937 of Tolkien's *The Hobbit*, which introduced the world to the wizardly figure of Gandalf.

Given the reading proclivities of the early computing community, it's not surprising that the word 'wizard' cropped up early in the digital context. By the start of the 1980s, it had begun to be widely used within Usenet groups to designate someone wise in the ways of computing: a 'system wizard', for example. The term soon became not so much slang as a title formally identifying someone with a profound understanding of a piece of software, upon whose powers one might call in an emergency.

Unlike self-designated 'hackers' – whose skills tended to be broad – the title of wizard denoted intricate knowledge of one particular area, and brought with it plenty of associated mystically inflected vocabulary. Performing 'heavy wizardry', for instance, meant taking on extremely complex practical tasks; while practising 'deep magic' meant exhibiting an incredible level of theoretical knowhow.[138]

As befitted their status, early computing wizards also tended to be people who were allowed to use systems – usually found, in the early 1980s, only within academic institutions – for purposes barred to ordinary users, such as pursuing their own coding interests. This led, by the end of the decade, to the addition of a so-called 'wizard mode' to some pieces of software, which once enabled allowed users to perform advanced tasks and reconfigure variables in a way normally not permitted.

# Wise web wizards

By the early 1990s, a new kind of computer wizardry was emerging that didn't require any human intervention at all. In 1991 Microsoft (among others) began introducing its customers to a helpful new kind of software it called a wizard, designed to take users through a particular task step-by-step. The most important such task was installing new software, but interactive guides were created for everything from drawing graphs and charts to word processing.[139]

It was the beginning of the end for human wizards as a mainstream computing phenomenon, not least because personal computing had simply outgrown the relatively small communities of early adoption – and manufacturers had begun to make usability by non-experts a central factor in design. By the 2000s, software wizards were almost universal, while its human meaning was in terminal decline: a minor parable of computing's trajectory from the private practice of a subset of initiates to its emergence as a mass phenomenon, where machine systems themselves provide 'magic' solutions to users' needs.

Today, you're unlikely to find many self-respecting programmers referring to themselves as wizards – at least outside the context of online gaming where, thanks in no small part to Tolkien's influence, wizards are among the most common of all classes of fantasy characters.

# 95.

# Disk drives

Originally, a disk drive was simply something that drove disks. Specifically, it was the machinery that rapidly rotated early computer disks — rather like extremely fast vinyl records — so that digital data could be written onto them, and read off them.

The word 'disk' itself was first used in a computing context towards the end of the 1940s. The word had been in circulation as a description of recording media since the 1880s — when it began to be used as a shorthand for the phrase 'phonograph disk' — and it was natural to extend it to one among a number of rival data storage mechanisms for computing.

It wasn't until 1953 that computer disks took the great leap forward into becoming the storage mechanism of choice for the medium. This decision came in the January of that year thanks to IBM researchers at its small research and development lab in San Jose, California. Specifically, it was the head of the lab, Reynold B. Johnson, who decided that the future of digital information lay in the form of

# Disk drives

Random Access Memory (RAM) recorded on Magnetic Disks; and that they would be the people to make this happen.[140]

Random Access Memory simply means memory that can be accessed in any order, picked at random – as opposed to memory that has to be read in the order in which it was recorded, like information on a cassette tape or vinyl record. This means that RAM can be used for storing temporary information that repeatedly needs to be overwritten while a computer is performing a task.

Johnson and his team duly conceived of a way of achieving this flexibility: a highly mobile read-write 'head' that would automatically be able both to record and read information from the surface of a spinning magnetic disk. All they needed was to build an integrated mechanism able both to spin the disks and to operate the read-write head: a 'disk drive'.

After three years of work the IBM team's product finally hit the market in September 1956, under the snappy title of the IBM 350 Random Access Method of Accounting and Control Disk File – or IBM 350 RAMAC for short. It was the world's first commercial disk drive, weighed over a ton, and had a capacity of five megabytes.

The meaning of the word 'drive' has shifted considerably, in computing terms, since those early days. Where once it described the machinery devoted physically to driving a computer disk, it has gradually become a shorthand for both the slot into which a disk is placed and any storage medium attached to a computer. Link your phone or a USB memory stick to a PC, for example, and once the machine has detected them they will appear as 'drives' ready for files to be copied, deleted, saved or searched.

One ancient habit – at least in computing terms – that has proved hard to shake is the tendency to label the main hard drive of a computer as the 'C:' drive. This dates back to the days of Microsoft's first Disk Operating System (DOS), in which the letters A and B were reserved by the system to be applied to 'floppy disks' (as early 8 and 5¼-inch removable disks were known) that might be inserted into the machine. One drive might be needed for a data disk, and one for a 'system' disk from which the computer could be booted. Thus the machine's internal hard disk got the letter 'C' – and, when CD-ROM drives later came to be added to computers, these tended to get the letter 'D'.

Even today, something akin to this letter-based allocation of drives remains on most machines. But it's only the drum of the hard disk that physically gets driven round in anything like the original sense any more, and even that may be on the way out thanks to the increasing popularity of immobile 'solid-state drives' – a phrase whose verbal self-contradictions are unlikely to bother most users.

# 96.

# Easter eggs

The tradition of Easter egg hunts, where chocolate and other treats are hidden for children to seek out, is at least in parts extremely ancient. Eggs were a pre-Christian symbol of rebirth and fertility across much of the world, and some authors trace back the tradition of children searching for hidden eggs at least to ancient Sumer – and subsequently to the worship of the Teutonic goddess Eostre, after whom the festival of Easter is named, and whose rites involved both eggs and rabbits.

Others note the beginning of the official White House Easter Egg Roll in 1878 (which put the stamp of state patronage on a popular pre-existing tradition) as a pivotal moment in promoting modern egg-related Easter activities.[141] When it comes to computing, however, 1979 is the year usually given for the arrival of quite another sense of 'Easter egg'.

This was the year in which the video game Adventure appeared for the Atari 2600 console – one of the first ever 'action adventure' games, in which players explored a fantasy world of castles and catacombs (i.e. different coloured blocks and blobs) while attempting to gather magical items. In those days, game designers didn't tend to get credited for their work so Adventure's creator, Joseph Warren Robinett, built into the game a secret extra task.

If players picked up a small, near-invisible object and brought it to a particular place in the game, they became able to walk through a normally solid wall and access a hidden area in which the words 'Created by Warren Robinett' were spelled out vertically. When the secret content was discovered after the game's release, Atari dubbed it an 'Easter egg' in honour of the tradition of searching for hidden goodies.[142]

The practice of building hidden details and special messages into products was, of course, nothing new: authors, broadcasters, film-makers and artists had long been including references, details and cameos that only a select audience would 'get'. The interactive nature of software, however, meant that something slightly different was now going on. As the very nature of a game like Adventure suggested, a software Easter egg was more like a hidden place or experience than merely a reference – something akin to an Easter egg physically hidden in a house or garden.

It's also a term, and a tradition, that has achieved enormously widespread use in the software field since 1979, with perhaps the most populous category of contemporary Easter egg consisting of amusing responses to selected user actions: a programming trick that's come to be regarded as a useful way of making software seem more 'human' and appealing.

Apple's voice-activated personal assistant, Siri, boasts a well-scripted range of Easter egg responses to questions ranging from 'I love you' (Answer: 'I hope you don't say that to all the mobile phones') and 'Will you marry me?' ('My EULA [End User Licence Agreement] does not cover marriage. Apologies.') to 'What's the meaning of life?' ('Forty-two.'). Similarly, Google's use of 'doodles' to mark important anniversaries – where the logo on its homepage is turned into an appropriate tribute, such as the centenary of Alan Turing's birth – is a kind of Easter egg, and often involves further hidden features (the Turing doodle, for example, functions as a 'live action Turing Machine with twelve interactive programming puzzles').[143]

Few modern Easter eggs, though, can compete for sheer oddity with a bonus feature coded within the 1995 edition of Microsoft's spreadsheet program Excel: a three-dimensional video game called 'Hall of tortured souls' accessed by executing a particular button combination in row 95 of a blank spreadsheet. The game itself contained hidden rooms and commands, as well as bizarre pictures on its virtual walls of staff at Microsoft: a hidden surprise that even the wildest imagination might not have expected.[144]

# 97.

# Microsoft family names

The origins of some internet names are more intriguing than they first appear, with the email service Hotmail being a case in point.

Launched commercially in 1996 by Sabeer Bhatia and Jack Smith, it was the world's first free web-based email service — a fact its creators marked by opening it up to the world on 4 July, American Independence Day, to symbolize the new freedom of being able to use email via the web anywhere in the world.

Having decided they wanted their company to include the world 'mail', Bhatia and Smith also wished to include the letters HTML in its title, signifying its relationship with the Hypertext Markup Language (HTML) that formed the basis of the world wide web. The name they came up with was thus originally written 'HoTMaiL' in order to emphasize HTML.

This typography was long gone by the time Microsoft bought Hotmail in December 1997 for a reported $400m, where it joined a family including some of computing's most famous names (at least until the announcement in August 2012 that the Hotmail brand would gradually be replaced by a new service, Outlook.com).[145]

'Microsoft' itself was named after a combination of the words 'microcomputer' and 'software', thanks to its founding in April 1975 when two young friends, Paul Allen and Bill Gates, saw an advert for the new Altair 8800 microcomputer and decided that there was a business opportunity in building a version of the programming language BASIC (Beginner's All-purpose Symbolic Instruction Code) for the system.

Originally hyphenated, 'Micro-soft', years later the company's two founders recalled toying with names like 'Outcorporated Inc and Unlimited Ltd', or simply 'Allen & Gates', before deciding that – in Allen's phrase – it needed to 'have a longevity and identity way beyond the founders'. As the company grew, a certain literalism was much in evidence in naming policy, as in the case of its first great success, 1981's Microsoft Disk-Operating System (MS-DOS).

Microsoft's second great success as a company after MS-DOS was almost called something equally functional: 'Interface Manager'. The company's head of marketing, however, managed to make the case that a slightly more seductive brand was required; and so it was decided to call its new Graphical User Interface 'Microsoft Windows' instead.

First announced in 1983, Microsoft Windows would go on to become the world's most-used operating system, bringing with it into the mainstream a host of terms and conventions. Among the less welcome of these popularizations is the 'Blue Screen of Death' – a widely used colloquial phrase for the blue screen that displays when a computer running a Windows-based operating system experiences a critical error.

Consisting of a blue background and white type, the BSOD today presents useful diagnostic information to users wanting to find out what might have caused their machines to go wrong. The very earliest versions of Windows, however, simply displayed an incomprehensible mess of code symbols – giving many users perhaps their first true taste of computing terror and confusion.

# 98.
# Why digital?

I've use the word 'digital' throughout this book in one of its most recent and vaguest senses: to describe technologies that involve digital data, or phenomena associated with these technologies. Yet the word itself is far older than such a definition suggests.

By the *OED*'s reckoning, the first written record of the word 'digital' appeared in English around 1425, describing 'a whole number less than ten'. Spelled *digitalle*, the word came from the Latin *digitalis*, meaning 'measuring the breadth of a finger' (today, the word *digitalis* is probably best known in English as a genus of the purple flower commonly called a Foxglove, its Latin name being a reference to the way its flowers seem perfectly sized to fit over the end of a human finger).

When was 'digital' used for the first time in the context of computing? By the 1940s, the term had begun to be used to describe discrete electronic values in electronics. It wasn't, however, until 1945 that the notion of an electronic digital computing machine was both born and named.

The specific machine in question was the Electronic Numerical Integrator And Computer (ENIAC), whose completion in 1946, at a cost of over six million dollars in today's money, gave the world its first electronic general-purpose computer. Thanks to a machine that weighed over 25 tonnes and needed 150 kW of power to operate, the age of digital computing had begun.

What really drove digital into the mainstream was not simply computing, but the steady development over the second half of the twentieth century of digital formats to replace the analogue storage media for everything from sound and images to video and text. Over the last two decades of the century, Compact Disks (better known as CDs) and Digital Versatile Disks (DVDs) largely replaced vinyl records and videos. Then, over the first decade of the twenty-first century, the direct distribution of digital files via the internet – known, inevitably, as 'digital distribution' – began to replace these storage media in turn, dispensing with everything other than the pure bytes of information itself.

Today, 'digital' is everywhere. We watch digital television; we analyse the digital economy; we speak about digital culture and digital trends. Increasingly, though, 'digital' seems at risk of becoming the victim of its own success. We live in a thoroughly digital age – which may well mean we stop needing the word at all. We no longer say 'digital computer'. Soon enough, we may no longer say 'digital' anything else.

# Why digital?

Except, of course, if we're talking about fingers or fox-
gloves: two sense of the word that are likely to remain
even when ones and zeros no longer need to be mentioned.

# 99.
# Filing away our data

Why do we keep data on our computers in 'files'? Like many other computing terms it began its life within inverted commas, as an analogy to existing information and organization systems – in this case, the analogy being a filing cabinet.

In use from the start of the 1950s, the notion of a computer 'file' initially described the physical medium on which information was stored – the medium, in the earliest computers, of punched cards rather than silicon disks.

It was only during the 1960s that computer files began to achieve their present sense, of virtual objects contained within computer storage media. Even then, however, the legacy of the word itself exerted a certain pressure on the medium, embodying the concept of each digital file as a distinct object, to be stored, copied and edited much like the physical items from which they took their name.

As computer scientists such as Jaron Lanier have pointed out, the very idea of using a 'filing system' to store data on a computer embodies limitations that are by no means inevitable in an electronic medium – and whose continuation owes something to the metaphorical force of the words first used to describe them.[146]

Modern computer files come in many different kinds, each performing different functions and matched to different kinds of software – a process of identification and matching in which file 'extensions' are crucial. Essentially a qualifying suffix appended after a full stop – the letters 'pdf' in a filename such as 'Book.pdf', for example – these extensions and their origins must rank among the most-used and least-understood aspects of digital vocabulary around. It's a mysteriousness abetted by the fact that they're usually only three letters long – a limitation that was once encoded in early file systems, but that today exists mainly because of inertia and compatibility issues.

Some file extensions have virtually entered into daily vocabulary, at least around the office. These include '.doc' (standing for 'document' and used in many word processing packages), '.xls' (used in Microsoft's Excel spreadsheet package, and standing for 'Excel spreadsheet') and '.pdf' (which stands for 'portable document format' and was created by Adobe in 1993 as a way of preserving the exact layout of a printable document across different computers).

Other extensions can seem obscure, but make considerably more sense once spelled out. A '.csv' file, for example, will consist of data in a 'comma-separated values' format: that is, a list of information separated by commas.

Similarly, understanding the nature of a '.gif' file becomes a little easier once you know it means a 'Graphics Interchange Format' file, designed to store a relatively simple image. By contrast, the '.jpeg' method of storing images stands for the 'Joint Photographic Experts Group', who created .jpeg files in order to establish an international standard for storing high quality photographs as computer files.

There are many hundreds of file types in the world today, some of which have even become words in their own right – '.MP3' for example, which is one of the world's most popular digital forms of audio encoding, and derives its name from the Moving Picture Experts Group, responsible for many of the world's standard audio and video formats.

Most of the time, we don't even notice they're there. And yet they form a vital layer of our verbal world – not least because it's these few letters that tell both us and our machines exactly what kind of information we're dealing with.

# 100.
# Artificial intelligence and Turing tests

Artificial Intelligence is a familiar enough idea today to have a earned an acronym – AI – that provided the title of a Steven Spielberg film in 2001. As a phrase, however, it's considerably older, having been coined in 1955 by the American computer scientist John McCarthy for a conference he and the cognitive scientist Marvin Minsky organized the following year at Dartmouth College.[147]

Interestingly, researchers during the first few decades of computing consistently tended to overestimate the potential of artificial intelligence, and of the rate at which machines would catch up with human cognition. In a sense, it wasn't until computer programmers tried to design machines capable of mimicking human intelligence that they discovered just how immensely complicated a phenomenon intelligence actually is.

Six decades later, AI has become less a radical claim about the future than a description of the way in which many everyday devices and applications are able to learn and adapt their behaviour. What we mean by the word 'intelligence' has, in parallel with other terms such as 'memory', expanded beyond the human – although few would yet claim that machine and human intelligence have much in common.

It's a distinction embodied in the formal division of artificial intelligence into 'strong' and 'weak' AI, where 'weak' denotes basic machine problem-solving and 'strong' denotes the as-yet-hypothetical matching of human intelligence by machines.

Of course, the degree to which machines can appear human was a subject of fascination since well before the era of computing; and it's one that was given its definitive contemporary form even before AI entered the scene: the Turing Test.

In 1950, the pioneer of modern computing Alan Turing published a paper entitled 'Computing Machinery and Intelligence'. In it he argued that actually determining whether or not a machine could 'think' was almost impossible, given the difficulty of defining thinking in the first place. Therefore he proposed a test to act as a proxy for this question: could a machine convince a person that it was, in fact, another human being?[148]

One of the most intriguing early developments in exploring Turing's proposition arrived in 1966, thanks to an early computer 'chatterbot' program called ELIZA (named after Eliza Doolittle in George Bernard Shaw's play *Pygmalion*) which was designed to 'talk' with users by a simple process of turning back their words into predetermined questions, in the style of a psychotherapist. The program's creator, MIT computer scientist Joseph Weizenbaum was shocked to discover that this crude simulacrum of empathy and interest was enough to fool some users into thinking of ELIZA as human. This phenomenon came to be known as the 'ELIZA effect', and is used today to describe the tendency to ascribe human motivations to computer effects.[149]

Anthropomorphism is nothing new. Pleasingly, though, there's also a term for the reverse of the ELIZA effect, born thanks to the founding in 1991 of the Loebner Prize for Artificial Intelligence – a now-annual enaction of the Turing Test under controlled conditions to try to find the world's most convincing conversational computer program.

The Loebner Prize asks participants to spend five minutes conversing via a computer console with an unseen other, who might either be human or a machine. While few programs, even today, can fool an experienced interlocutor for long, the organizers of the prize noticed that some real people were rated during the test as 'almost certainly a machine'. It's a phenomenon that has been dubbed 'the confederate effect' as it involves one of the human 'confederates' failing to prove their humanity. This was observed during the very first Prize, in 1991, when one human participant's detailed knowledge of Shakespeare was deemed too wide-ranging and precise to be anything other than mechanical.[150]

# . . . and finally

This is a book dedicated to new words and meanings – and the shifting senses of older words, whose stories within language can never end as long as they remain in use.

Consider the word I've used to describe what you're reading at this very moment: a 'book'. In Old English, in the form of *boc*, it originally meant any written document; a term derived in turn from the ancient Germanic word for a beech tree, *bokiz*, which provided the smooth, soft bark onto which runic writing was inscribed.

Physicality, then, is an intimate aspect of the history of the book: the power and possibility that came with turning language into a physical object, fit to be kept for generations, bearing words across time and space in a way no speaker could hope to match.

Today, written words shorn of physical restrictions are travelling in ways inconceivable less than a century ago. The relationship between written language and the mere physical stuff of the world – tablets, trees, wood, paper – has been decisively broken.

## . . . and finally

So why write a 'book' at all? Why bother crafting a text that aims to be a complete, self-contained object when it's now possible to enter into the open-ended arenas of digital words; to update, respond and interact as you encounter your audience?

One answer, I hope, is implicit in the kind of book I have attempted to write here. In a digital age, written knowledge is almost infinitely to be found online, together with enough discussion and debate for several lifetimes. We do not read a book, today, simply to learn. Rather, we turn to crafted, bounded texts looking for words that aspire to permanence.

With them, we can perhaps conjure a quality of time and space that more than ever we deserve in our lives – a place for turning mere information into something that belongs to us.

Thank you for reading.

# Acknowledgements

A thousand thanks first of all to my agent, Jon Elek, and my editor, Richard Milner at Quercus, who between them made this book possible. It's the kind of book I've spent most of my life wanting to write, and I'm extremely grateful to be given the chance. As ever, my wonderful wife, Cat, has also helped to keep me sane, and looked over pretty much everything I've demanded she read. As a reader, I was lent a copy of Bill Bryson's *Made in America: An Informal History of the English Language in the United States* when just entering my teens, and it became the first of many books about words I delightedly absorbed many of them the loving gifts of my mother and Cacti, who taught me to care deeply about language. Most recently, Mark Forsyth's *The Etymologicon* and his blog The Inky Fool, at http://blog.inkyfool.com/ have been an inspiration and source of considerable pleasure. For someone who spends a considerable amount of time looking up words online, it would be perverse not to acknowledge the marvels of Wikipedia as a boundless source of ideas,

# Acknowledgements

interconnections and references; or, indeed, the sheer usefulness of Google Books. Both have been invaluable, together with all the online apparatus any author now takes for granted, from search engines and electronic texts to the cloud-based word systems in which I store most of my work. In a more specialized vein, the Language Log blog occupies an important place in my digital reading life, as does Douglas Harper's Online Etymological Dictionary. In print, the eminent academic and linguist David Crystal also deserves particular thanks, both as a communicator of big ideas around language and as an early investigator of what he has termed 'netspeak'. A few of the ideas in this book first featured in a short-lived column for *Prospect* magazine, 'Tom's Words', which gave me my first taste of etymological publication. The editor at the time, David Goodhart, also did much to help me become a better writer and to shake off tergiversatory academic habits. Other sources of ideas and inspiration – and there are many – are covered in the notes and further reading sections, and I am grateful to them all. All errors, needless to say, are my own; or are likely to stem from my too trustingly repeating a mistaken source, something of which there's no shortage online. Please let me know about any howlers via Twitter at @TomChatfield and I'll try to get them fixed in future editions.

# Select bibliography and further reading

This list combines those resources I've found most useful in writing this book with those I'd recommend for further etymological exploration.

It's a deliberately selective and idiosyncratic collection. For a more complete list of sources and citations, see the notes in the next section.

## Books

Aitchison, Jean, *Language Change: Progress or Decay?* (Cambridge UP, 2000)

Baron, Naomi, *Always On: Language in an Online and Mobile World* (OUP USA, 2010)

Crystal, David, *The Stories of English* (Penguin, 2005)

Crystal, David, *Txtng: The Gr8 Db8* (Oxford UP, 2009)

Crystal, David, *Language and the Internet* (Cambridge UP, 2006)

Deutscher, Guy, *Through the Language Glass: Why The World Looks Different In Other Languages* (Arrow, 2011)

# Select bibliography and further reading

Forsyth, Mark, *The Etymologicon: A Circular Stroll through the Hidden Connections of the English Language* (Icon, 2011)

Pinker, Steven, *The Stuff of Thought: Language as a Window into Human Nature* (Penguin, 2008)

## Online

The Eggcorn Database – http://eggcorns.lascribe.net/

English Language & Usage at Stack Exchange – http://english.stackexchange.com/

Google Books – http://books.google.com/

Google Ngram Viewer – http://books.google.com/ngrams

The Inky Fool – http://blog.inkyfool.com/

Internet Gaming Slang glossary – http://www.ling.lancs.ac.uk/staff/tony/onlineslang.htm

Know Your Meme – http://knowyourmeme.com/

Language Hat – http://www.languagehat.com/

Language Log – http://languagelog.ldc.upenn.edu/nll/

Merriam-Webster's Open Dictionary – http://www3.merriam-webster.com/opendictionary/index.php

The New Hacker's Dictionary (the Jargon File) – http://www.outpost9.com/reference/jargon/jargon_toc.html

The OED's Oxford Words blog – http://blog.oxforddictionaries.com/

The Online Etymological Dictionary – http://www.etmonline.com/

Schott's Vocab (no longer updated) – http://schott.blogs.nytimes.com/

Shady Characters, the secret life of punctuation —
http://www.shadycharacters.co.uk/

Snopes — http://www.snopes.com/

The Urban Dictionary — http://www.urbandictionary.com/

The Virtual Linguist — http://virtuallinguist.typepad.com/
the_virtual_linguist/

Wikipedia (natch) — http://en.wikipedia.org/wiki/
Main_Page

Word Spy — http://www.wordspy.com/

# Notes and references

1   Jeff Howe, 'The rise of crowdsourcing', *Wired*, June 2006, online at http://www.wired.com/wired/archive/14.06/crowds.html

2   For a full account of the 'Apple key' story, see Andy Hertzfeld (2004), *Revolution in The Valley: The Insanely Great Story of How the Mac Was Made*, under the heading 'Swedish Campground', p. 152.

3   The first appearance in print of both 'hypertext' and the related notion of 'hypermedia' came in Ted Nelson (1965), 'Complex information processing: a file structure for the complex, the changing and the indeterminate', online at http://dl.acm.org/citation.cfm?id=806036

4   A full text of the original letter is online, courtesy of the Liberty Fund, at http://oll.libertyfund.org/?option=com_ staticxt&staticfile=show.php%3Ftitle=666&chapter=81892& layout=html&Itemid=27

5   The earliest reference to Voltaire's letter using the 'kindle' translation that I can locate in English comes in Louis Mayeul Chaudon's 1786 *Historical and critical memoirs of the life and writings of M. de Voltaire*, a translation of Dom Chaudon's French original, which on page 348 renders the end of Voltaire's letter on Swift thus: 'He [Voltaire] looked on

everything as imitation. The most original writers, said he, borrowed from one another. Boyardo had imitated Pulci, and Ariosto Boyardo. The instruction we find in books is like fire; we fetch it from our neighbours, kindle it at home, communicate it to others, and it becomes the property of all.' The line may then have been popularized via a review in the November 1786 issue of the *Gentleman's Magazine*, which quotes the above extract in full (although it doesn't think much of the book, which is dismissed as a 'laboured panegyric').

6     This exchange about the origins of computer daemons is reported in full online at http://ei.cs.vt.edu/~history/Daemon.html

7     Ward Cunningham's full 'correspondence on the etymology of wiki' is online at http://c2.com/doc/etymology.html

8     See Raph Koster's own account of the nature of nerfing on his website at http://www.raphkoster.com/gaming/nerfing.shtml

9     Further details for all metric units and prefixes can be browsed on the official website of the international bureau for weights and measures at http://www.bipm.org/en/si/si_brochure/chapter3/prefixes.html

10     Although Tuvalu apparently doesn't earn as much from its domain as it thinks it should: see 'Internet domain riches fail to arrive in Tuvalu', *Independent on Sunday*, 18 July 2010, online at http://www.independent.co.uk/news/world/australasia/internet-domain-riches-fail-to-arrive-in-tuvalu2029221.html

11     See Adam Fox, *Oral and Literate Culture in England, 1500–1700* (Clarendon Press, 2000), p. 56.

12     See Dennis G. Jerz's account of the play and the word 'robot' at http://jerz.setonhill.edu/resources/RUR/

13     Asimov's 'Runaround' is collected in *I, Robot* (Voyager, 2004),

# Notes and references

with the first mention of 'robotics' coming on page 38: 'But then, advances in robotics these days were tremendous . . .'

14 The only mention of 'cyberspace' in 'Burning Chrome' comes in its enigmatic third sentence: 'I knew every chip in Bobby's simulator by heart; it looked like your workaday Ono-Sendai VII, the "Cyberspace Seven," but I'd rebuilt it so many times that you'd have had a hard time finding a square millimeter of factory circuitry in all that silicon.' The full text is online at http://www.acsu.buffalo.edu/~rrojas/BurningChrome.htm

15 In *Neuromancer*, 'cyberspace' is most iconically glossed in this context: 'Case was twenty-four. At twenty-two, he'd been a cowboy, a rustler, one of the best in the Sprawl. He'd been trained by the best, by McCoy Pauley and Bobby Quine, legends in the biz. He'd operated on an almost permanent adrenaline high, a byproduct of youth and proficiency, jacked into a customized cyberspace deck that projected his disembodied consciousness into the consensual hallucination known as the matrix.' See William Gibson, *Neuromancer* (Voyager, 1995), pp. 11–12.

16 A full account of these and other additions to the *OED* in March 2011 – from which the quotation in this paragraph comes – can be found online at http://public.oed.com/the-oed-today/recent-updates-to-the-oed/previous-updates/march-2011-update/

17 See Neal Stephenson, *Snow Crash* (Penguin, 2011), p. 33: 'He is not seeing real people, of course . . . The people are pieces of software called avatars. They are the audiovisual bodies that people use to communicate with each other in the Metaverse.'

18 This passage can be found on page 192 of the '30th anniversary' 2006 paperback edition of *The Selfish Gene* (OUP), as well as being endlessly quoted online.

19    Heinlein's explanation of 'grok' comes on p. 322 of the 2007 paperback edition of *Stranger in a Strange Land* (Hodder).

20    As reported in the *Guardian*'s account of 13 June 2011, online at http://www.guardian.co.uk/world/2011/jun/13/gay-girl damascus-tom-macmaster

21    The original European Parliament paper is online at http://www.europarl.europa.eu/RegData/etudes/etudes/cult/2002/318587/DG-4-CULT_ET(2002)318587_EN.pdf

22    The story was originally reported in the *Gainesville Times* on 1 March 2012, online at http://www.gainesvilletimes.com/archives/63990/

23    For further details, see Declan McCullagh's article for CNET News from 23 April 2004 at http://news.cnet.com/2100-1032_3-5198125.html

24    The *Washington Post* was one of the first to pick up the Tyson Gay story, in Mary Ann Akers's blog 'The Sleuth' on 1 July 2008 – see http://voices.washingtonpost.com/sleuth/2008/07/christian_sites_ban_on_g_word.html

25    Letter of 13 November 1878. For a full account of Edison's struggles around the electric light at this time, including the quotation in question, see Robert Freidel and Paul B. Israel *Edison's Electric Light* (John Hopkins UP, 2010), pp. 19–23.

26    The original log book, complete with original bug, can be found in the National Museum of American History; see http://americanhistory.si.edu/collections/object.cfm?key=35&objkey=30 for details and a selection of images.

27    Perhaps the definitive account of all such bugs can be found within the iconic hacker's dictionary *The Jargon File*, a version of which is maintained by Eric S. Raymond at http://www.catb.org/jargon/ – or which can be purchased in paperback form as *The New Hacker's Dictionary*.

# Notes and references

28    Within the game-world of EVE Online, debating how many carebears it takes to feed a griefer really is a genuine, and statistically serious, question, as the debate at https://forums.eveonline.com/default.aspx?g=posts&m=347596 illustrates.

29    Taken from p. 237 of the 1978 Hutchinson edition of one of Burgess's most unusual books, *1985*.

30    Explore the Ngram Viewer for yourself at http://books.google.com/ngrams

31    The full text of *Through the Looking Glass* can be found online at http://www.gutenberg.org/files/12/12-h/12-h.htm including this passage introducing the idea of portmanteau words, in which the character of Humpty Dumpty is speaking to Alice: "'Well, 'SLITHY' means 'lithe and slimy'. 'Lithe' is the same as 'active'. You see it's like a portmanteau—there are two meanings packed up into one word.'"

32    See Linda Stone's own explanation of Continuous Partial Attention at http://lindastone.net/qa/continuous-partial-attention/

33    To be precise, the original coining of the term 'Streisand Effect' was on 5 January 2005 in this piece for Techdirt by Mike Masnick, the blog's editor – http://www.techdirt.com/articles/20050105/0132239.shtml

34    See the *OED*'s March 2012 update, in which 'super-injunction' was admitted to the dictionary for the first time, detailed online at http://public.oed.com/the-oed-today/recent-updates-to-the-oed/previous-updates/june-2012/a-sublime-and-superlative-quarter-of-contrasts/

35    See Alan Rusbridger's 'The Trafigura fiasco tears up the textbook', first published on the *Guardian* website on 14 October 2009 – http://www.guardian.co.uk/commentisfree/liberty-central/2009/oct/14/trafigura-fiasco-tears-up-textbook

36  A complete text of *Three Men in a Boat* can be found online at http://www.gutenberg.org/files/308/308-h/308-h.htm

37  See http://research.microsoft.com/apps/pubs/default.aspx?id=76529 for the full paper.

38  See Marc D. Feldman, Maureen Bibby and Susan D. Crites, '"Virtual" Factitious Disorders and Munchausen by Proxy', *Western Journal of Medicine*, June 1998, vol. 168, no. 6, online at http://www.ncbi.nlm.nih.gov/pmc/articles/PMC1305082/pdf/westjmed00333-0055.pdf

39  A thorough account of the Kaycee Nicole case and several other famous fictitious online stories can be found at http://www.snopes.com/inboxer/hoaxes/kaycee.asp

40  See Ben Hammersley, 'Audible revolution', *Guardian*, 12 February 2004, online at http://www.guardian.co.uk/media/2004/feb/12/broadcasting.digitalmedia

41  Detailed information on Gelernter's early work with Eric Freeman on lifestreaming can be found at http://cs-www.cs.yale.edu/homes/freeman/lifestreams.html while perhaps the earliest citation for the term 'lifestream' itself is an article for the *Washington Post* of 3 April 1994 called 'The cyber-road not taken', partly reproduced online at http://www.wordspy.com/words/lifestreaming.asp

42  Philip Steadman, *The Evolution of Designs: Biological Analogy* (Routledge, 2008), p. 260.

43  Manfred E. Clynes and Nathan S. Kline, 'Cyborgs and space', *Astronautics*, September 1960 – online at http://cyberneticzoo.com/wp-content/uploads/2012/01/cyborgs-Astronautics-sep1960.pdf

44  A full text of Vinge's essay can be found online at http://www.rohan.sdsu.edu/faculty/vinge/misc/singularity.html

45  An excerpt from Kurzweil's book, including the quoted passage,

can be found online at http://www.npr.org/books/titles/
138051148/the-singularity-is-near-when-humans-transcend-
biology#excerpt

46   Google's full official history of itself can be read online at
     http://www.google.co.uk/about/company/history/

47   The exchange takes place on page 120 of the 1995 Heinemann
     edition of *The Hitchhiker's Guide to the Galaxy: A Trilogy in
     Five Parts.*

48   Alain de Botton, *Status Anxiety* (Penguin, 2005), p. 3.

49   Those wishing to know everything there is to know about the
     Pwnies should visit http://pwnies.com/

50   Google officially labels this the 'hacker' language version of
     its interface: see https://www.google.com/webhp?hl=xx-hacker

51   Scott Fahlman's own in-depth account of his invention of the
     emoticon is online at http://www.cs.cmu.edu/~sef/sefSmiley.
     htm and concludes with the honest observation that 'I probably
     was not the first person ever to type these three letters in
     sequence, perhaps even with the meaning of "I'm just kidding"
     and perhaps even online. But I do believe that my 1982
     suggestion was the one that finally took hold, spread around
     the world, and spawned thousands of variations . . .'

52   A scan of the 1881 page from *Puck* magazine is online at
     http://upload.wikimedia.org/wikipedia/commons/e/ee/
     Emoticons_Puck_1881_with_Text.png

53   The full Word Spy article on 'slacktivism' is online at http://
     www.wordspy.com/words/slacktivism.asp

54   A partial transcript of the key 'pajamas' exchange was published
     on 17 September 2004 on the Fox News website at http://
     www.foxnews.com/story/0,2933,132494,00.html

55   One of the earliest usage I've found of 'gamification' is from
     a 2009 'Loyalty Expo' in Hollywood, Florida, as reported in

this July 2009 summary at http://www.scribd.com/doc/ 17718638/Loyalty-Expo-2009-in-Review

56  See Steve Mann's 2002 reference page 'sousveillance' at http://wearcam.org/sousveillance.htm – elements of which, including the first definition of 'sousveillance' itself, were first published in March 2002 in the magazine *Metal and Flesh* under the title 'Sousveillance, not just surveillance, in response to terrorism', online at http://wearcam.org/metalandflesh.htm

57  See Ian Kerr's blog post on 'equiveillance' and subsequent comments at http://www.anonequity.org/weblog/archives/ 2006/01/exploring_equiv_1.php

58  Ron Rosenbaum's original article 'Secrets of the Little Blue Box' appeared in *Esquire* in October 1971; a full scan is online at http://www.historyofphonephreaking.org/docs/ rosebaum1971.pdf

59  The full text of that iconic first spam email is online at http://www.templetons.com/brad/spamreact.html – note that, just to make things even worse, the entire message was written in capital letters.

60  A full text of Wolfe's 1965 essay on McLuhan is online at http://www.digitallantern.net/mcluhan/course/spring96/wolfe. html

61  It's extremely difficult to find a source for Newton Love's origination of the term 'CamelCase' other than versions of the story cited on Wikipedia at http://en.wikipedia.org/wiki/ CamelCase thanks to the vanishing of the original 1995 online discussion at http://users.sluug.org/~newt/

62  The full entry is online at http://www.bradlands.com/weblog/ september_10_1999/

63  The department of linguistics at Lancaster University has a useful glossary of online gaming slang, featuring these and

many more terms, at http://www.ling.lancs.ac.uk/staff/tony/onlineslang.htm

64 See Bartle's email exchange about the origins of 'mob', reproduced at http://mud.wikia.com/wiki/Mob

65 Hilbert's 1904 paper was the text of a lecture given at the third international congress of mathematicians at Heidelberg, 'On the Foundations of Logic and Arithmetic', and the idea of proving the internal consistency of mathematics occupied much of his energy over the following decades – until, in 1931, Kurt Gödel's seminal 'incompleteness theorem' formally demonstrated that Hilbert's aims were impossible.

66 The Hofstadter cartoon is online at http://xkcd.com/917/

67 Perhaps inevitably, the 'talk' section accompanying Wikipedia's own explanation of its TL;DR policy is itself packed with accusations that the article is too long to read; see http://en.wikipedia.org/wiki/Wikipedia_talk:Too_long;_didn't_read

68 Know Your Meme offers a usefully illustrated summary of RTFM's first appearance in print at http://knowyourmeme.com/memes/rtfm

69 See Harry McCracken's 'Fanboy! The strange true story of the tech world's favourite putdown', posted on 17 May 2010 on the Technologizer blog, at http://technologizer.com/2010/05/17/fanboy/

70 See the Guild's official YouTube channel at http://www.youtube.com/user/watchtheguild

71 The Syndicate's official website is http://www.llts.org/

72 The 'Picard facepalm' is helpfully reproduced, alongside other miscellaneous info, at http://knowyourmeme.com/memes/facepalm

73 For a marvellous, detailed history of the Turk and its times, see Tom Standage's book *The Mechanical Turk: The True Story*

*of the Chess-playing Machine That Fooled the World* (Penguin, 2003).

74  Explore Amazon's mechanical Turk service online at https://www.mturk.com/mturk/welcome

75  The full text of Cory Doctorow's *For the Win* is available via his website at http://craphound.com/ftw/Cory_Doctorow__For_the_Win.htm under a Creative Commons Attribution-Non Commercial-ShareAlike 3.0 licence.

76  A useful history of geocaching, as well of information on how to get involved, is online at http://www.geocaching.com/about/history.aspx

77  See http://www.dartmoorletterboxing.org/history%20of%20Letterboxing.htm for more details on Dartmoor letterboxing, including the contents of its first ever instance in 1854, when 'a Dartmoor guide named James Perrott placed a glass bottle at Cranmere Pool, and encouraged hikers that made the considerable walk to the site to leave a calling card as a record of their achievement . . .'

78  You can listen to the 'song of the grass mud horse' and read further details from the *China Digital Times* at http://chinadigitaltimes.net/2009/02/music-video-the-song-of-the-grass-dirt-horse/

79  See 'Phrases for lazy writers in kit form' by Geoffrey Pullum, posted on 27 October 2003 on Language Log, at http://itre.cis.upenn.edu/~myl/languagelog/archives/000061.html

80  See 'Snowclones: lexicographical dating to the second' by Geoffrey Pullum, posted on 16 January 2004 on Language Log, at http://itre.cis.upenn.edu/~myl/languagelog/archives/000350.html

81  See 'Egg corns: folk etymology, malapropism, mondegreen, ???' by Mark Liberman, posted on 23 September 2003 on Language

Log, at http://itre.cis.upenn.edu/~myl/languagelog/archives/ 000018.html and updated with Pullum's suggestion of 'egg corn' as a term on 30 September. For an exhaustive list of eggcorns, the Eggcorn Database at http://eggcorns.lascribe.net/ is recommended reading.

82    The original 'Jargon Watch' column from March 1995 is online at http://www.wired.com/wired/archive/3.03/jargon_watch. html

83    Of all his work, Andrew Keen's book *Digital Vertigo: How Today's Online Social Revolution Is Dividing, Diminishing, and Disorienting Us* (Constable, 2012) explores these ideas most closely.

84    John Markoff, 'Pools of Memory, Waves of Dispute', *New York Times*, 29 January 1992, online at http://www.nytimes.com/ 1992/01/29/business/business-technology-pools-of-memory-waves-of-dispute.html

85    See Irving Biederman and E. A. Vessel, 'Perceptual Pleasure and the Brain', *American Scientist*, vol. 94, May–June 2006, online at http://geon.usc.edu/~biederman/publications/ Biederman_Vessel_2006.pdf

86    See 'Infovores & Ignotarians', 25 July 2006, on The Personal Genome, at http://thepersonalgenome.com/2006/07/ infovores_ignot/

87    Maggie Jones's essay 'Shutting themselves in', published in the *New York Times* on 15 January 2006, is still one of the best places to begin learning about *hikikomori*, and is online at http://www.nytimes.com/2006/01/15/magazine/15japanese. html?pagewanted=all

88    See the BBC's report 'Australian dies after 'planking' on balcony, police say', from 15 May 2011, at http://www.bbc. co.uk/news/world-asia-pacific-13389207

89   One example among many, uploaded on 11 June 2011, is the video 'cone-ing at a McDonalds – epic reaction', which as of September 2012 had a cool 1.3m YouTube views – see http://www.youtube.com/watch?v=78DBf2_9Esg

90   The site http://www.horsemanning.com/ includes historical photos from the 1920s and an extensive guide to the modern revival of the trend (that is, lots of silly photos).

91   See http://www.unfriendfinder.com/

92   Word Spy puts the earliest citation for 'meatspace' at 21 February 1993 in the 'Austin Cyberspace Journal Newsletter', and I am unable to find an earlier citation; see http://www.word spy.com/words/meatspace.asp

93   See page 175 of David Gerrold's *When Harlie Was One* (Ballantine, 1972), in which one character asks, 'You know what a virus is, don't you? . . . The virus program does the same thing.'

94   See Freeman John Dyson, 'Search for Artificial Stellar Sources of Infrared Radiation', *Science*, vol. 131, 3 June 1960, online at http://www.islandone.org/LEOBiblio/SETI1.HTM

95   Robert Bradbury died at the age of 54 in March 2011. A detailed tribute to the man and his work, including the matrioshka brain concept, is online at http://www.sentientdevelop-ments.com/2011/03/remembering-robert-bradbury.html

96   Greg Bear first introduced the Taylor algorithms in his 1988 novel *Eternity*, explaining that they allow 'programs to completely determine the nature of their systems. Thus, a downloaded mentality could tell whether or not it had been downloaded . . .'

97   To be precise, the phrase 'commencing interweb link' appeared approximately 14 minutes and 30 seconds into 'Eyes', the sixteenth episode of season one of *Babylon 5*, 'Eyes' – first

broadcast on 13 July 1994. See http://www.imdb.com/
title/tt0517650/

98    A full transcript of the third debate between George W. Bush
      and John Kerry is online at http://www.debates.org/index.php?
      page=october-17-2000-debate-transcript – while Bush's im-
      mortal deployments of 'internets' came in this context: 'Parents
      are teaching their children right from wrong, and the message
      oftentimes gets undermined by the popular culture . . . We
      can have filters on Internets where public money is spent.
      There ought to be filters in public libraries and filters in public
      schools so if kids get on the Internet, there is not going to be
      pornography or violence coming in.'

99    See George P. Krapp, *The English Language in America*, first
      published in 1925.

100   See Sergey Brin and Lawrence Page, 'The Anatomy of a Large-
      Scale Hypertextual Web Search Engine', online at http://
      infolab.stanford.edu/~backrub/google.html

101   See *The Software Age* by Paul Niquette, online at http://www.
      niquette.com/books/softword/part0.htm

102   For more details, see the main Tor Project site at https://
      www.torproject.org/

103   For a flavour of Silk Road's operations, see this story from 7
      August 2012 on Ars Technica, under the headline 'Study
      estimates $2 million a month in Bitcoin drug sales' – http://
      arstechnica.com/tech-policy/2012/08/study-estimates-2-
      million-a-month-in-bitcoin-drug-sales/

104   See Vinton Cerf, Yogen Dalal and Carl Sunshine, 'Specifications
      of internet transmission control program', December 1974 –
      online at http://tools.ietf.org/html/rfc675 – whose use of the
      term 'internet' followed on from the term 'internetworking'
      in papers such as Vinton Cerf and Robert Kahn, 'A Protocol

for Packet Network Intercommunication', May 1974 – see http://www.cs.princeton.edu/courses/archive/fall06/cos561/papers/cerf74.pdf

105   See Tim Berners-Lee, *Weaving the Web: The Past, Present and Future of the World Wide Web by its Inventor* (Harper, San Francisco, 1999) – the naming discussion comes on p. 23 of this edition.

106   See http://www.st-isidore.org/ for full details of the Order of Saint Isidore of Seville's online activities.

107   Explore the Pope2you website at your leisure at http://www.pope2you.net/ and make up your own mind.

108   As reported on the Vatican's own news site at http://www.news.va/en/news/follow-the-pope-on-twitter-for-lent

109   Walter Isaacson, *Steve Jobs: The Exclusive Biography* (Little, Brown, 2011), p. 63.

110   'Steve Jobs Almost Named The iMac The MacMan, Until This Guy Stopped Him' – an extract from Ken Segall's *Insanely Simple* (Portfolio, 2012) – tells the story in full at http://www.fastcodesign.com/1669924/steve-jobs-almost-named-the-imac-the-macman-until-this-guy-stopped-him

111   For a detailed history of Linux and its naming, see Ragib Hasan's 'history of Linux' at https://netfiles.uiuc.edu/rhasan/linux/

112   Jennifer Viegas, *Pierre Omidyar: The Founder of Ebay* (Rosen, 2006), p. 52.

113   See W. K. English, D. C. Englebart and Bonnie Huddart, 'Computer-aided Display Control Final Report', 1 July 1965, online at http://archive.org/details/nasa_techdoc_19660020914

114   The episode, 'Lisa's wedding', was first broadcast on 26 March 1995; a full script can be read online at http://www.snpp.com/

episodes/2F15.html including the immortal line: 'Bart: [pause] Meh.'

115　There's a detailed and useful discussion of 'meh' on the Language Hat blog from 13 April 2007, online at http://www.languagehat.com/archives/002716.php

116　See 'Word featured in The Simpsons becomes latest addition to Collins English Dictionary', *Daily Telegraph*, 16 November 2008, online at http://www.telegraph.co.uk/news/3467717/Word-featured-in-The-Simpsons-becomes-latest-addition-to-Collins-English-Dictionary.html

117　Sam Leith, '"Meh" is more useful than "weaselnose"', *Daily Telegraph*, 17 November 2008, online at http://www.telegraph.co.uk/comment/personal-view/3563562/Meh-is-more-useful-than-weaselnose.html

118　The original exchange, dated 19 October 2004, remains online in the comments section at http://onepamop.livejournal.com/240305.html?thread=2071217#t2071217 (and is long, obscene and often extremely hard to follow).

119　Wikipedia has a scan of the image at http://en.wikipedia.org/wiki/File:Newsweek_preved.jpg

120　Wikipedia also features an entry dedicated to 'bear surprise', complete with an image of the original painting, at http://en.wikipedia.org/wiki/Bear_surprise

121　Werner Buchholz, 'The System Design of the IBM Type 701 Computer', first published October 1953 in the *Proceedings of the I.R.E.*, vol. 41, no.10, online at http://ieeexplore.ieee.org

122　This discussion on Word Origins is one of several places to tell the Andy Williams story about the origins of cookies http://wordoriginsorg.yuku.com/topic/6654#.UF9wso1mSIo

123　Lou Montulli lies behind quite a few other innovations too. As he puts it on his own website at http://www.montulli.org/lou

'I'm largely to blame for several innovations on the web, including cookies, the blink tag, server push and client pull, HTTP proxying, proxy authentication, HTTP byte ranges, HTTPS over SSL, and encouraging the implementation of animated GIFS into the browser . . .'

124 The Information Commission Office's guide to EU cookie law is online at http://www.ico.gov.uk/for_organisations/ privacy_and_electronic_communications/the_guide/cookies. aspx

125 Marc Prensky's 'Digital Natives, Digital Immigrants' – originally published in *On the Horizon* (MCB University Press, vol. 9, no. 5, October 2001) – can be read along with many of his other writings via his website at http://www.marcprensky. com/writing/

126 The original Intel memo on 'Netiquette guidelines', dated October 1995, can be read in full online at http://www.ietf. org/rfc/rfc1855.txt

127 For an extremely thorough discussion of Donkey Kong naming debates, see http://www.snopes.com/business/misxlate/ donkeykong.asp

128 The Technologizer blog published a usefully detailed piece on Mario and his namesake, 'The true face of Mario,' by Benj Edwards on 25 April 2010, online at http://technologizer. com/2010/04/25/mario/

129 See Chris Kohler's interview with the creator of Pac-Man, Toru Iwatani, for *Wired* on 21 May 2010 – thirty years after the game's release – at http://www.wired.com/gamelife/2010/05/ pac-man-30-years/

130 Bill Wasik offered an in-depth account of his flashmob escapades in the March 2006 issue of *Harper's Magazine*, online at http://www.harpers.org/archive/2006/03/0080963

# Notes and references

131 Godwin's original post is reproduced online at http://w2.eff.org/Net_culture/Folklore/Humor/godwins.law together with several notable subsequent corollaries and comments.

132 Mike Godwin, 'Meme, counter-meme', *Wired*, October 1994, online at http://www.wired.com/wired/archive/2.10/godwin. if_pr.html

133 Satisfactory sources are difficult to find in this area, with most relying on anecdote and personal recollection – see the discussion at http://english.stackexchange.com/questions/ 40013/whats-the-origin-of-beta-to-describe-a-user-testing-phase-of-computer-devel. According to the Jargon File, the terms were first used at IBM only in the 1960s – http://catb.org/jargon/html/B/beta.html

134 See http://en.wikipedia.org/wiki/Gender_of_connectors_and_ fasteners for further details.

135 CNN reported the event on 26 November 2003 under the headline '"Master" and "slave" computer labels unacceptable, officials say' – see http://edition.cnn.com/2003/TECH/ptech/ 11/26/master.term.reut/

136 One of the earliest citations for 'mother board' is a 1956 article in the journal *Electronic industries,* vol. 15, issue 2, which explains how, 'In order to create complete equipment, several . . . wafer-like subassemblies must be connected to a mother board . . .' See also the discussion at http://english.stackexchange.com/questions/10386/why-motherboard-is-used-to-refer-to-main-board-of-computer which dates the first occurrence of 'daughter board' in print to 1965.

137 For a staggeringly detailed history of the pilcrow, see Keith Houston's three-part history of this punctuation mark on his Shady Characters site, which begins at http://www.shady characters.co.uk/2011/02/the-pilcrow-part-1/

138 *The New Hacker's Dictionary* has details on all these wizardly

terms, and several others, at http://www.outpost9.com/ reference/jargon/jargon_38.html#TAG1994 including this gem: '*wave a dead chicken /v./* To perform a ritual in the direction of crashed software or hardware that one believes to be futile but is nevertheless necessary so that others are satisfied that an appropriate degree of effort has been expended . . .'

139 Microsoft dates its own first software 'wizard' to the launch of Publisher 1.0 in 1991; see the Microsoft News article 'For 10 Years, Microsoft Publisher Helps Small Business Users "Do More Than They Thought They Could"' from 15 October 2001, online at http://www.microsoft.com/en-us/news/features/ 2001/oct01/10-15publisher.aspx

140 The story of the development of the first disk drive is told in fully illustrated detail in a pamphlet by the Magnetic Disk Heritage Centre entitled 'The IBM350 RAMAC Disk File: Designated An International Historic Landmark by The American Society of Mechanical Engineers', first published on 27 February 1984. Running to fifteen pages, it can be read online in full at http://www.magneticdiskheritagecenter.org/ MDHC/RAMACBrochure.pdf

141 The White House Easter Egg Roll even has its own official White House webpage at http://www.whitehousehistory org/whha_press/index.php/backgrounders/white-house-easter-egg-roll/

142 Wikipedia's summary of the first video game Easter egg is probably the best online account of this seminal digital event – see http://en.wikipedia.org/wiki/Adventure_(Atari_2600)

143 A 'museum' of every single Google doodle can be explored online at http://www.google.com/doodles/finder/2012/ All%20doodles

144 Full details of Excel's utterly bizarre 1995 Easter egg, complete

with pictures, are online at http://eeggs.com/items/719.html

145   Dean Wilson, 'Microsoft to replace Hotmail with Outlook.com',
2 August 2012 – http://vr-zone.com/articles/microsoft-to-
replace-hotmail-with-outlook.com/16890.html

146   See Jaron Lanier, *You Are Not a Gadget: A Manifesto* (Allen
Lane, 2010), and his discussion, beginning on p. 12, of the
'locked-in' notion of the computer file as 'a set of philosophical
ideas made into eternal flesh'.

147   The original 'proposal for the Dartmouth Summer research
project on artificial intelligence' is dated 31 August 1955 with
four authors cited: J. McCarthy, M. L. Minsky, N. Rochester,
and C. E. Shannon, and is online at http://www-formal.
stanford.edu/jmc/history/dartmouth/dartmouth.html

148   Alan Turing, 'Computing Machinery and Intelligence', *Mind*,
October 1950, vol. 59, no. 236, pp. 433–60, online at http://mind.
oxfordjournals.org/content/LIX/236/433

149   You can 'talk' to ELIZA – and several other chatbots – online
at http://nlp-addiction.com/eliza/

150   Full details of the Loebner Prize are on its website at
http://www.loebner.net/Prizef/loebner-prize.html while a
brilliant account of taking part in the competition, and its
philosophical implications, can be found in Brian Christian's
*The Most Human Human: What Artificial Intelligence Teaches
Us about Being Alive* (Viking, 2011) – including the cited story
of Shakespeare expert Cynthia Clay, who in November 1991
was deemed a computer by three different judges on the grounds
that 'no one knows that much about Shakespeare'.